BASIC WOODWORKING & CARPENTRY ... with projects

BY JERROLD R. CLIFFORD

TAB BOOKS Inc.
BLUE RIDGE SUMMIT, PA. 17214

FIRST EDITION

FIRST PRINTING—APRIL 1980

Printed in the United States of America

Library of Congress Cataloging in Publication Data

Clifford, Jerrold R.
Basic woodworking & carpentry.

Includes index.
1. Carpentry—Amateurs' manuals. 2. Woodwork—Amateurs' manuals. I. Title.
TH5606.C55 694 79-25287
ISBN 0-8306-9884-1
ISBN 0-8306-1058-8 pbk.

Cover photo courtesy of Georgia Pacific.

To Sarah Clifford
a fine author
a wonderful mother
a loving memory.

Contents

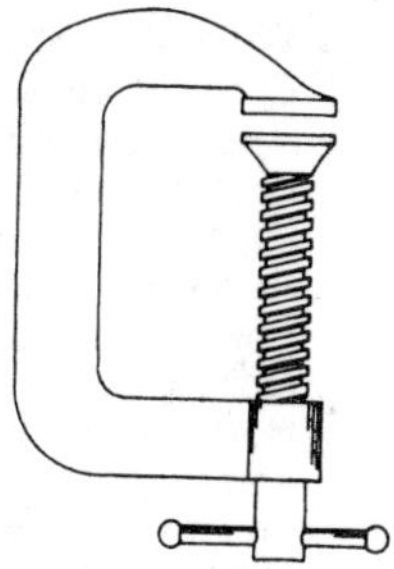

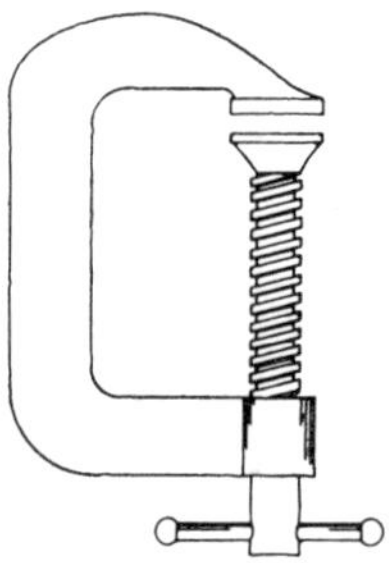

Introduction

Carpentry is building and repairing items—especially those made of wood. Since many things in an apartment or home, and perhaps the home itself, are made of wood, the ability to work with wood—to shape, mold, cut, repair, and finish it is certainly a useful skill to develop. The techniques described in this book can be applied to a variety of craft projects . . . from creating a wall ornament to constructing the wall to put it on. A large amount of knowledge or equipment is not necessary for beginning this rewarding craft. Whether you perform a small job such as repairing a loose cabinet door or a large job such as laying a wooden floor, you can get the kind of satisfaction that can only be obtained from doing the work yourself.

People often shy away from carpentry projects because they feel they do not have the coordination and dexterity required for the successful completion of the jobs. Actually, there are many carpentry projects that can be accomplished easily with little skill. As you gain the experience to tackle larger tasks, your confidence will increase. Remember, even the most skilled professional carpenter was inexperienced at one time.

Jerrold R. Clifford

1

Anyone Can Construct Carpentry Projects

Although the use of simple calculations is sometimes necessary, proficiency in mathematics is not a requirement for carpentry crafts. You will not need above average strength or other unusual physical attributes or abilities. What is necessary are the things most people already possess—common sense, reasoning ability, ability to follow directions and diagrams and a desire for accomplishment. Age is not a restriction. Youngsters and senior citizens can both enjoy this craft. You will not need a large amount of money to get started. It is possible to spend a considerable sum on carpentry crafts, but it really isn't necessary. And once you have started, maintenance costs are quite reasonable.

Many carpentry projects do not require a large working area. Don't be afraid to begin a craft because you live in a small apartment or house where space is at a premium. And don't be discouraged if you lack experience in the use of tools. I will explain how to select the proper tool for the job and explain how to use it.

WHAT IT TAKES TO GET STARTED

The first thing you will need is a set of old clothes. These should be loose enough to allow complete freedom of motion but fit snuggly enough to avoid bunching up when you bend over. A pair of comfortable shoes is a must. They should afford maximum protection against falling objects such as a screwdriver that slips or the hammer you accidentally drop—sandals and slippers should be avoided. The shoes should also have a solid heel and sole because stepping on a nail or wooden splinter in your bare feet or with a flimsy shoe will definitely take the fun out of carpentry.

A shop apron affords additional protection. An old barbeque apron that ties securely will also do the job. When you are working on a ceiling or wall or other overhead object, wearing a cap is a good idea. An old baseball cap will generally serve the purpose.

Avoid wearing scarves, jewelry, and ties. Loose items, particularly around the throat, are dangerous around tools. Try to wear clothing that has at least one pocket because it is a handy storage place for a pencil and a tape measure. Speaking of pencils, a mechanical pencil with a pocket clip is not only the most convenient but it is also safest. The pencil will be available when you need it and you will not stab yourself on a sharp point. A self-winding tape measure is a practical item. It can fit into a pocket and be available at all times for making measurements.

ESTABLISHING A WORK AREA

It is both inconvenient and annoying to search for a special tool or plan before you start work on your project. Considerable time can be wasted as well. Establishing a well planned work area can not only help you avoid aggravation, but also increase your enjoyment of carpentry crafts.

Considerable space is not a requirement, but the area should be large enough for you to work comfortably. Often, a portion of a basement or a garage can be used. However, these areas can often be poor choices. While affording the obvious advantages of allowing you to make noise without disturbing other members of the household and letting you be a little messy at times without worrying about disturbing your living environment, these areas also have several disadvantages associated with them. Often, they are unheated so working on projects in cold weather is uncomfortable. Some basements have a tendency to be damp. Not only will the dampness decrease your enjoyment of carpentry, but the moisture can have an adverse affect on your finished projects. Basements sometimes contain an old furnace in need of cleaning or repair that gives off smoke that is not only unhealthy to breathe but is also annoying when you are concentrating on a difficult procedure. When a garage is used, it is often necessary to compete for space with cars, lawn mowers, garden tools, bicycles and items that have been placed there for storage. Garage lighting that is usually fine for assisting you when you are entering or leaving your car is not the best lighting in which to work.

Choose an area that will be comfortable for you to be in for extended periods. It should be dry and not too cool in winter or too warm in summer. Make sure that you have good lighting so that your plans and the markings on wood can be easily seen. The working environment should be a relaxing one. If you enjoy listening to music, a radio tuned to your favorite station would add to your enjoyment of carpentry crafts. If you would rather be involved in your own thoughts, create an atmosphere that would encourage this.

The work area should have an adequate electrical supply. Several outlets, preferably grounded to provide an additional safety factor, are extremely convenient when you need to connect an extra lamp or portable power tool. You can easily tell if your outlets are grounded. The receptacle will have three openings—two rectangular parallel slots and a rounded opening just below and between them. Ungrounded outlets have only two openings. To

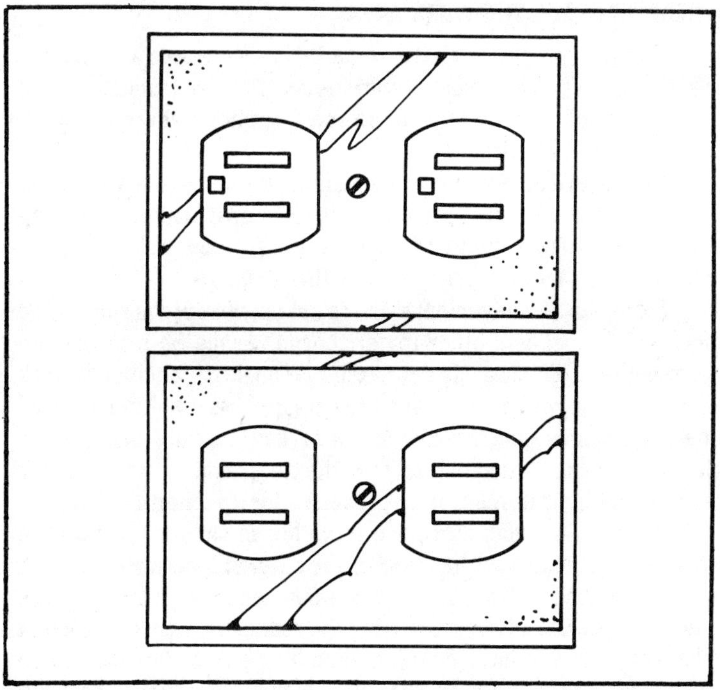

Fig. 1-1. At the top is a pair of grounded outlets. Notice that they have three openings. Grounded outlets provide an extra safety feature and are generally found in most newer homes and apartments. At the bottom are ungrounded outlets.

determine which type you have, compare your outlets to those shown in Fig. 1-1.

Good ventilation is a must. Breathing vapors from wet paints, varnishes, and other chemicals in a closed room can seriously affect your health. Be sure the area you select has a window or a fan to help circulate air. Besides making the work area safer, windows and fans provide an extra measure of comfort.

Once you have selected the work area, there are two items needed to furnish it—a strong sturdy table or workbench and enough storage space for tools and parts. Although many home handymen and professional carpenters prefer to build their own workbenches and cabinets, it really isn't necessary. Often, an old wooden table and discarded cabinets or other sturdy furniture will serve admirably. Once you have some familiarity with woodworking and basic tools, constructing a solid workbench is easy. You will find detailed plans for this project in chapter 7.

WHERE TO PLACE THE WORK AREA

Now that you know what is required in a work area and some of the advantages and pitfalls of various locations, consider several convenient spots around the home and apartment where the work area could be placed.

Basement. Many home work shops are set up in basements. Perhaps this is because basements can be messy without resulting in an untidy house. Additionally, the use of basement shops does not detract from the main living area of the home.

Garages. A convenient place to set up work areas since large pieces of lumber and other materials can usually be brought into them without difficulty. If the garage is small, an additional work area can be provided by moving the car if necessary. Even in small one-car garages, extra storage space can be built using extra shelves or cabinets that extend over the car's hood. Loft storage for ladders and light items can be created under pitched roofs.

Attics. Although attics are often hot in summer and do not provide a great deal of headroom, they can still be used as work areas. Installing an attic fan serves a dual purpose. It reduces the attic temperature while also cooling the rest of the house. Although many attics do not have flooring, sheathing-grade plywood can be placed over the supports to serve as a floor. Covering the floor with indoor-outdoor carpeting can muffle noise and provide some additional insulation against heat loss. Usually, fold-away staircases can be easily installed in the attic opening resulting in a convenient access to the work area. While a very attractive work shop can be created in an attic, do not use this portion of your home if you can't stand comfortably in it.

Closets. A closet can offer considerable tool storage space and a small walk-in closet can be converted to an attractive work area. This is particularly convenient for apartment dwellers who do not want to provide their landlords with permanent cabinets or to face the problem of dismantling a shop when they move. A little ingenuity can be used to create hanging shelves on the inside of the closet door. A folding work table can be used as a workbench. Hand tools can be stored on peg board suspended from a bar normally used to hang clothes. You can use a few loose bricks or heavy paperweights to keep the peg board from moving.

Spare rooms. If you are fortunate enough to have an extra bedroom, enclosed porch, or other room that is not being utilized, you have the potential for an excellent work area. Chances are that such rooms already have adequate lighting and ventilation along with suitable electric outlets. However, tracking sawdust into other rooms could prove to be an inconvenience. If you enjoy working during hours while others in the household are sleeping or watching television, you must consider the amount of noise you will make unless the room is well soundproofed.

Laundry Room. Often, laundry rooms can be used for carpentry work in addition to its intended purpose of cleaning and ironing clothes. By using drop cloths to protect appliances from dust and using a sturdy folding table, the laundry room can also be used as an excellent work area. However, due to space limitations you will probably have to clear the work table after each use and remove any dust before doing the laundry.

Anywhere. Although you should first gain some experience in the use of wood and tools before attempting such a project, furniture can be constructed that converts to a self-contained work area. These items are excellent for use where space is limited. It will allow you to work practically anywhere in the home or apartment.

WHAT YOU SHOULD KNOW BEFORE YOU START

Carpentry, like other crafts, assumes a familiarity with certain words and expressions. Although these terms will be examined in greater detail in later chapters, it is a good idea to gain some familiarity with them at this time.

Perhaps the most familiar item used in carpentry is the *nail*. Nails are available in several different shapes, sizes, lengths and widths. Each type is specifically designed for its intended purpose. The top part of the nail is called the *head*. Some nails have wide heads and notches for maximum holding power. Others have narrow heads to allow for fine work and to permit them to be hidden so that the heads won't be visible when your project is completed. Their primary purpose is to fasten two pieces of wood together. The tool used to accomplish this is called a *hammer*.

Another device used for fastening is the *screw*. Like nails, screws also come in various lengths and widths. They resemble nails but have a metal thread running along the length and a slot in the head. Screws are used to increase holding power or to fasten metal, plastic and other non-wooden objects. *Screwdrivers* are used to position screws.

Bolts resemble certain types of screws. However, they are generally sturdier, are longer or thicker than ordinary screws and usually do not have slotted heads. They are used in conjunction with *nuts* that twist onto the head and provide the holding power. *Lock washers* are often used with this combination to prevent the nuts from moving. Bolts are used when greater holding is needed, extra strength is desired or when the parts being fastened will be moving.

Staples are fasteners that are becoming increasingly popular with do-it-yourselfers and hobbyists. Resembling the familiar office paper fastener, they are generally sturdier and are ideal for quickly installing certain types of paneling and ceilings and for other decorating projects. Their big advantage is that they are easy to use. They can easily be placed with one hand using a *staple gun*.

Nuts, bolts, nails, screws, staples and similar items are referred to as *hardware*. Selecting the proper hardware for the job and obtaining needed items are described in chapter 4.

SAVE MONEY BY DOING YOUR OWN CARPENTRY

Anyone who has ever shopped for wooden shelves, cabinets, furniture, picture frames and similar items knows those items are expensive. Often these items are not exactly what you had in mind but you must accept what is available. One of the biggest expenses with regard to these items is labor. While skilled craftsmen could be required for difficult jobs, you can do many projects yourself and save money. There is a one-time cost for tools you will need but they can be used on many projects. You have the opportunity to select the wood and components you prefer and you gain the pride and satisfaction of doing the work yourself.

When you consturct the items yourself you can be sure of the quality that is built in. Often, items which are purchased in a store are not quite what they seem to be and prove a disappointment once they are in your home. Frequently these products do not last as long as you would like and so they soon have to be replaced.

Most people are well aware of the purchase price of their selections. However, few realize the total cost they pay. This is because they do not consider the many hidden expenses involved in obtaining their final choices. Your favorite store might not always have what you want, making it necessary to make calls or to travel from one store to another in hope of obtaining what you want. This costs you money in gasoline and wear on your car. If you are fortunate enough to find exactly what you want you could still be faced with the problem of how to get it home. This involves either the use of more gasoline or the payment of a delivery charge—either as an add-on charge or as part of the purchase price. Sometimes lumber yards do not charge extra for delivery. When they do, the charge is often only nominal. This will save you money on "hidden" costs.

CARPENTRY IS FUN

Carpentry can save you money, provide the satisfaction of creativity and offer an excellent, rewarding hobby. But most of all it's fun. And that's what crafts are all about.

2 Tools You Will Need

Since all carpentry craft projects involve the use of tools, it's a good idea to know something about them—their purpose, construction and use—before you plan your first job. By using the proper tool at the right time you will save considerable energy, avoid mistakes and obtain more professional looking results.

THE FIVE TOOL CATEGORIES

Different tools are designed for specific purposes and they can be divided into five distinct categories:

—Fastening
—Drilling
—Cutting
—Layout and gauging
—Clamping

TOOLS FOR FASTENING

A basic part of carpentry is the ability to fasten pieces of wood together. There are several ways to accomplish this. The most familiar technique makes use of a tool called a *claw hammer.* This tool got its name from the slotted metal portion that somewhat resembles the claws of an animal. There are two main categories of claw hammers—*curved claws* and *ripping claws* (Fig. 2-1). The curved type has a rounded claw which offers increased leverage

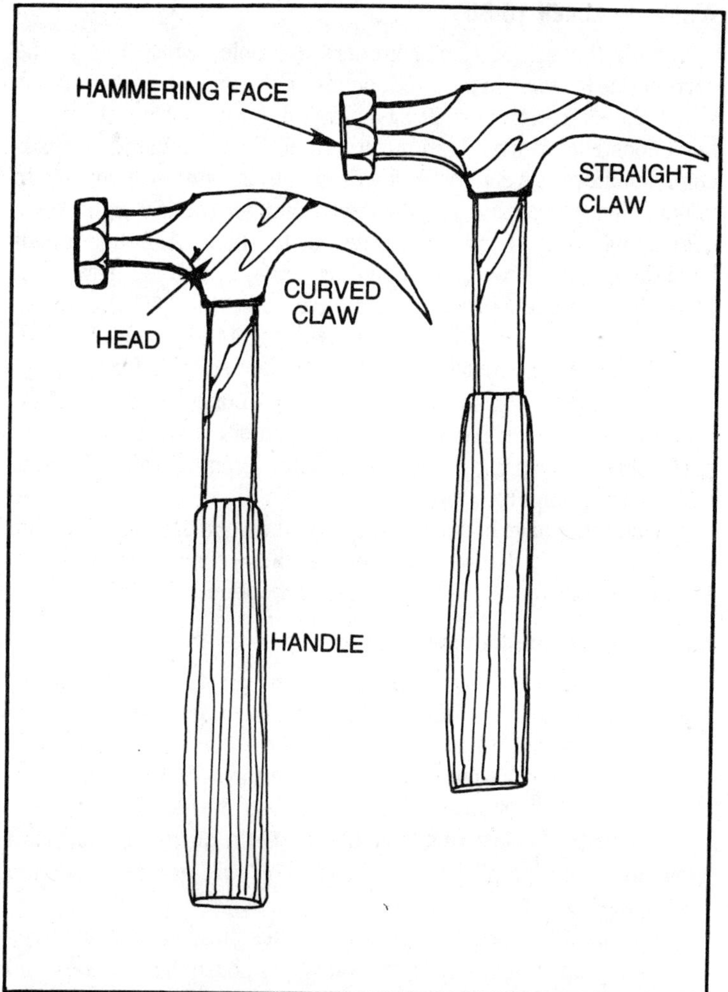

Fig. 2-1. The curved claw hammer shown on the left is designed to remove nails as well as drive them. The straight claws of the hammer on the right aids in pulling apart pieces of wood.

intended for removing nails. The straight, or ripping, claw is designed for separating pieces of wood that are already fastened together.

If you examine different claw hammers you will notice that some have faces that are flat while others are made slightly rounded or convex. The convex type offers the advantage of permitting you to drive a nail flush with the surface of your wood without damaging it. A mesh-type face is intended for use for rough framing.

WHICH HAMMER TO BUY

With the selection of hammers available, which one is the correct one for you? In part, it depends on the type of work you wish to do. Select a hammer that has a forged, hardened and tempered head constructed from high quality steel. Because hammer heads come in different weights ranging from 5 to 20 ounces, you have to make a choice here too. Try to select a hammer with a weight that is comfortable, but at the same time not too light. Although a light hammer might be more attractive to you in the store, using light hammers sometimes results in greater muscle fatigue than when heavier ones are used. Most people prefer a 14 to 16-ounce head for finishing work. The heavier weights are more suitable for framing.

The lengths of hammers also vary. Long handles provide greater swing leverage than short hammers and are therefore preferable for framing. Short handles are intended for use during the finishing stage of a project.

Hammers have handles made of either wood or metal. Wooden handles generally do not last as long as those constructed of steel, but some individuals prefer the feel of wooden handles.

How To Determine Hammer Quality

While hammers might look alike, their quality can vary considerably. Some of the features that comprise a quality product are easy to see, while others are not. Be sure that the hammer you select has a sharp V-shaped groove in the claw. This will allow you to remove small or headless nails. If you decide on a wooden handled model, be sure that the end grain of the hickory handle has a grain that runs parallel to the head. This will ensure maximum strength.

Quality hammers have smooth finishes on all striking surfaces and on the top surfaces of the claws. Other polished areas serve no purpose other than to improve appearance.

Since most people are not metal experts, the quality of the metal in a hammer head can't be determined by a sight inspection. Check for a label, tag, or carton description that states the materials used. The head on a quality hammer is made of drop-forged steel and has hardened striking surfaces. The metal in the body portion is softer so that metal fatigue is reduced.

How To Hammer Properly

Misusing hammers results in bent nails, marred wood and painful bruises. Learning how to use a hammer properly takes only a little time and it is easy. Get accustomed to the following method:

- —Hold the handle near its end.
- —Always hold the hammer firmly. This will keep it from slipping and allow for greater accuracy.
- —Use a full arm stroke to hit the nail squarely with the center of the hammer face. Hitting a nail at an angle will bend it.

How To Drive A Nail

Nails are used in conjunction with hammers to fasten pieces of wood together. Although the use of hammers is somewhat obvious, using them properly requires a little practice. If you have never driven a nail before, take a piece of scrap wood and try the following technique:

- —Place the tip of the nail against the wood and hold it between your thumb and forefinger near the head. This will allow the hammer to push your fingers away rather than giving them painful bruises.
- —Tap the head of the nail very lightly a few times with the hammer to make a slight indentation in the wood.
- —Hold the nail at a slight angle. Positioning nails in this manner provides stronger fastening than when they are driven straight into the wood.
- —Tap the nail several times with the hammer so that the wood is gripping the nail and holding it in position. This will allow you to remove your fingers from the nail so you will not bang them with the hammer.
- —Using firm, hard strokes hit the head of the nail with the hammer until the head of the nail is against the surface of the wood.

Whenever you plan on painting or staining wood, it is a good idea to drive the nail a little deeper into the wood so that the nail head is slightly below the surface. If you did this using only a

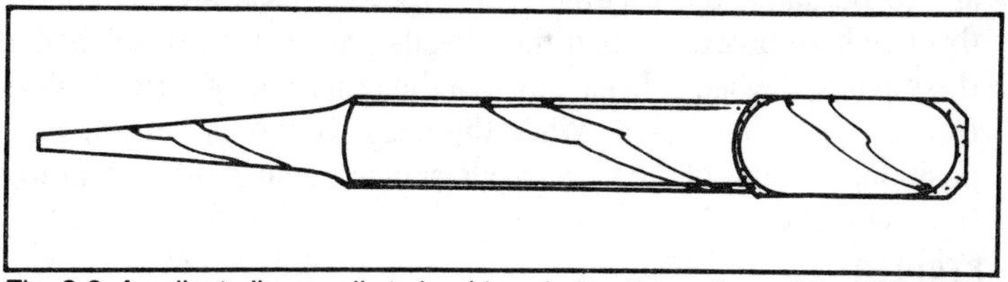

Fig. 2-2. A nailset allows nails to be driven below the surface of the wood. The small end is placed against the head of a nail which has been driven into the wood and the broad end is struck by the hammer.

hammer, the wood would be marred by the hammer itself. A device called a *nailset* (Fig. 2-2) prevents the marring of wood surfaces. A nailset is a cone-shaped device. Its tapered end is placed against the head of the nail and its broader end is struck by the hammer. This keeps the hammer head well away from the wood's surface.

How To Remove A Nail

Even expert carpenters occasionally make a mistake when driving nails or they are faced with a situation where a nail must be removed from a piece of wood. If this happens to you, the following technique will allow you to remove the nail while at the same time guarding against damage to the wood.

—Place a flat stick between the hammerhead and the surface of the wood. If the nail being removed is long or proves difficult to pull, use a thicker piece of wood to gain greater leverage.

—The claw of the hammer should have a V-shaped groove. Wedge it around the nail's shank between the head of the nail and the wood.

—Rock the hammer away from the nail. For most nails move the hammer backwards. For difficult nails, rocking the hammer sideways might keep the size of the nail hole small.

Special Nailing Techniques

When performing certain projects it isn't always possible to drive a nail directly from the flat surface of one piece of wood into the end of another. In such cases, the nail must be positioned at an angle through the side of one board and pass into the surface of the other. This technique is called *toenailing* and it is often used in carpentry projects. When you must toenail it is important to select the proper angle to drive the nail since it must penetrate well into both pieces of wood to insure that it is properly secured.

Another practice, called *clinching*, is used when two boards are to be laminated together. Several long nails are hammered through both boards so that their heads are securely positioned in the top board while at least 1 inch of the other end of the nail sticks through the other board. While the heads of the nails are kept in position, the ends of the nails sticking out are bent in the direction of the wood grain.

STAPLERS

Stapling can make some carpentry craft projects easier and faster to complete. Staplers are generally used with one hand and

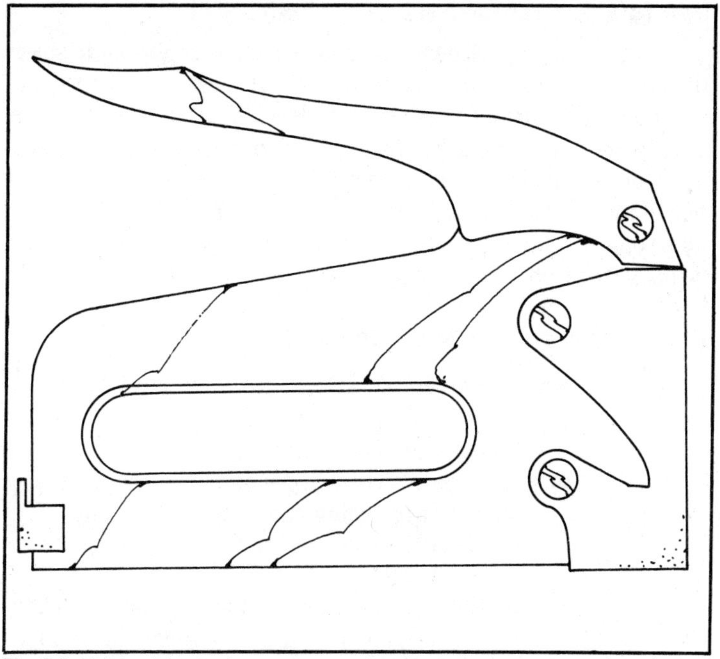

Fig. 2-3. Lightweight staplers are convenient for many carpentry craft projects.

are squeezed or slammed against the surface being fastened. Since only one hand is usually required for their use, the other hand can be used to position the material being stapled and to hold it in place. These staplers are inexpensive and can usually be purchased in hardware stores or the hardware section of many department stores.

Other types of staplers are used in the construction of homes and buildings. These are generally pneumatically or electrically driven. Because of their high cost, they are best left to professionals. Figure 2-3 shows a typical lightweight stapler.

SCREWDRIVERS

After the hammer, screwdrivers are the most frequently used tool in carpentry. This important tool is available in many shapes and sizes. The type to use depends on the screw you wish to position.

Ordinary slotted screwdrivers are intended for use with screws that have a straight slot. Phillips screwdrivers are used when the screw head has a crossed slot. Since the sizes of these slots vary with the screw being used, you will want to start with a small selection of three to four screwdrivers.

How To Select Screwdrivers

When making a carpentry craft project, it is important to select the proper screwdriver. Using a screwdriver that is the wrong size often results in the head of the screw developing burrs. Sometimes a screwdriver that is improperly matched to the screw head will slip. This could result in the wood's surface being damaged. Because the tips of the screwdrivers must properly fit the slot of the screw and be strong enough to turn the screw, very cheap screwdrivers often do not get the job done since they are generally made of soft metal.

How To Purchase The Proper Screwdriver

- Buy several screwdrivers rather than just one. Be sure that they are different lengths and sizes and that at least one is a Phillips.
- Since long screwdrivers permit the application of more power than a comparable short one and is less likely to lean, be sure that one of your screwdrivers is long.
- Purchase a short or stubby screwdriver. These are extremely useful for working in cramped spaces.
- If possible, select screwdrivers that are magnetic or that have a wedge or spring clip. These features will assist you in holding the screw for proper positioning.
- Some screwdrivers come with rubber or cushioned grips. These handles allow you to have a firmer grip on the tool, turn the screw more easily and they protect your hands from blisters.

The *offset* is another type of screwdriver. These are generally about 5 inches long and resemble the letter Z. One end of the offset screwdriver has a blade that is parallel to the handle. The other end has a blade at right angles. Because of the lever nature of this tool, it has several times the turning power of an ordinary screwdriver and is very helpful in turning stubborn screws. Because of their compact size, offset screwdrivers are ideal for working in tight spaces. The two blades allow a screw to be turned a small distance and then they can be switched to allow for additional small rotations. If you wish to limit the amount you initially spend on tools, wait until you gain experience in the use of screwdrivers before you purchase this type.

Spiral ratchet screwdrivers are often used by professional carpenters to speed the time used to drive screws. These screwdrivers are considerably more expensive than the standard screwdrivers and require some experience in driving screws if they are to be used properly. They are not recommended for people just starting to learn carpentry.

How To Drive Screws

- Create a pilot hole for the screw. This can be done by tapping a nail, which has a smaller diameter than the screw, several times with a hammer and then pulling it out. If you prefer, a small hole can be drilled. The actual length and width of these starter holes depends on the size of the screw to be driven and the type of wood used.
  - —For hard wood, make holes that are about as wide as the distance across the threaded part of the screw measured in the valley between the threads.
  - —In soft wood, make the hole about two-thirds as wide as the distance across the threaded part of the screw, again measured in the valley between the threads.
  - —Use smaller holes when placing screws in the ends of pieces of wood.
  - —If the screws being used are long, make holes about as long as the screw itself.
- Position the screw within the pilot hole.
- Holding the screw with your fingers or the wedge or clip feature of the screwdriver, turn the screwdriver in a clockwise direction.

Special Screwdriver Techniques

Despite the use of a pilot hole, it is still sometimes difficult to drive a screw. Rubbing a small amount of soap or wax on the threads of the screw sometimes makes it easier to drive.

Theoretically, removing a screw is simple. Just turn the screwdriver in a counterclockwise direction. However, sometimes screws are stubborn and you could experience some difficulty in turning it. If this happens to you, try heating the screw head slightly. The tip of a soldering iron is ideal for this if you happen to have one. If you don't, try heating the tip of a worn or cheap screwdriver over a stove and then touch it to the screw head. Once the screw has been heated, allow it to cool and then try to remove it with the screwdriver.

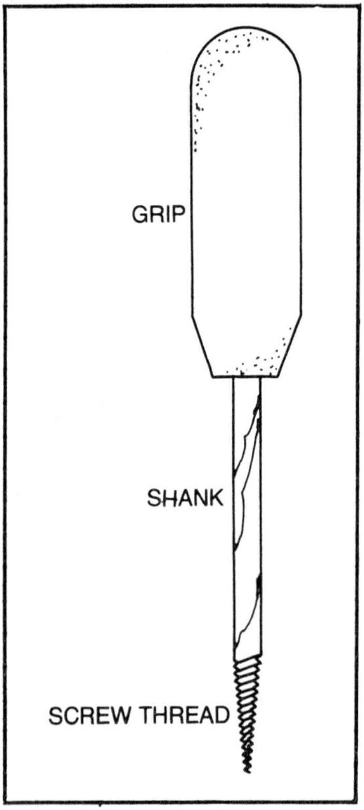

Fig. 2-4. A screw starter is often helpful when you want to position screws.

When positioning screws, it can sometimes be annoying to have the screwdriver constantly slip from the slot in the head of the screw. If you find this happening to you, tape the two items together with a small piece of tape. Once the screw is turned in the wood a few times, you can remove the tape.

If you are using a good screwdriver and find that the screw slot tends to burr, you might have to replace the screw. If you have a similar screw made of steel, try driving it first and then remove it and position the softer screw.

A *screw starter* (Fig. 2-4) is a tool that looks as if someone welded a screw and a screwdriver together. As its name implies, this device is often useful for making a starter hole before positioning the actual screw. However, care must be taken to make sure that the hole created is not bigger than the screw to be positioned. To use this tool, make a small pilot hole. Turn the handle of the screw starter in a clockwise direction while at the same time pushing it into the hole.

PLIERS

A pair of pliers (Fig. 2-5) is one of the most useful tools in carpentry. They range in size and strength from small jeweler's types to large varieties used in plumbing and heavy duty wiring. All-purpose adjustable pliers cost only $3 to $6 and they are a must for every work shop. They can be helpful for a variety of purposes—pliers can be used to twist wire, assist in gripping objects, remove broken nails or screws from wood or in conjunction with a wrench to place bolts.

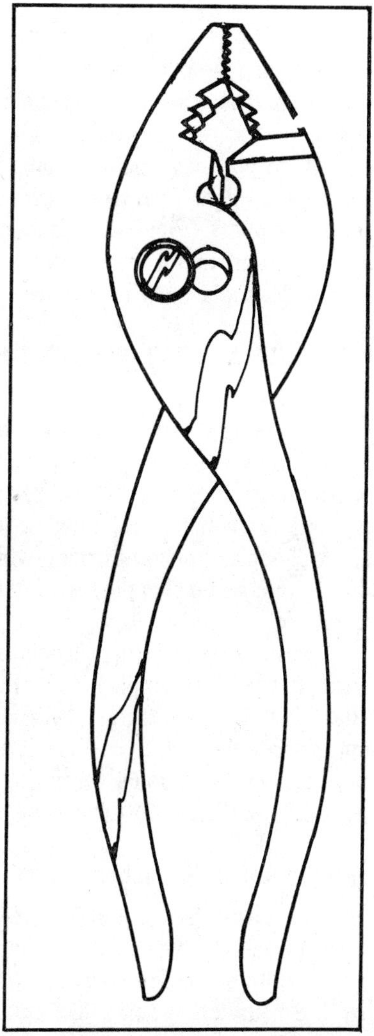

Fig. 2-5. A pair of all-purpose adjustable pliers is an important tool for any workshop.

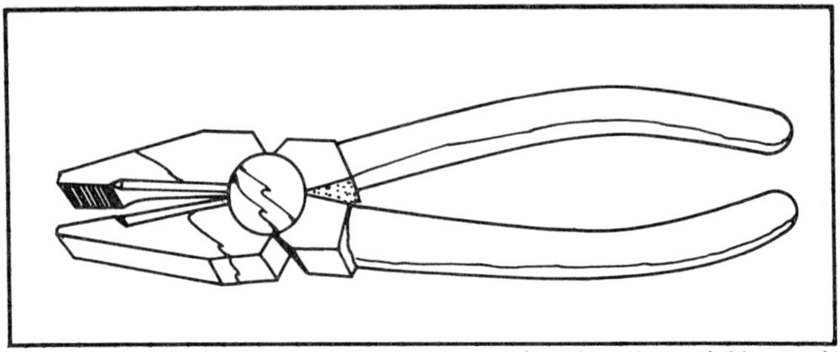

Fig. 2-6. Lineman's pliers assist in bending and cutting wire and thin metal strips.

All-purpose pliers often come with plain metal handles. Inexpensive plastic grips are available that slide onto these handles and help keep the tool from slipping in your hand.

Once you have gained some familiarity with the all-purpose pliers, you might wish to add other types to your work shop. The lineman's pliers (Fig. 2-6) is an aid in bending and cutting wire or strips of thin metal, while the long nosed variety (Fig. 2-7) is invaluable in places where there isn't very much room to work.

WRENCHES AND TROUBLESHOOTING TOOLS

Wrenches are the types of tools that really aren't required for most carpentry craft projects. They often aren't necessary until something goes wrong! They are used to enhance muscle power when turning a nut onto a bolt or whenever extra turning leverage is needed. A wrench can be used instead of a pair of pliers to hold nuts and to keep them from turning while a screw is positioned. This is often done when two pieces of metal or thick boards are fastened together.

Sometimes a screw will be stubborn and will not turn. A useful trick that you can employ if you have a screwdriver with a square shank is to position the screwdriver in the screw head. Fasten the appropriate wrench to the shank of the screwdriver and rotate the wrench. The additional turning power supplied by the wrench is generally sufficient to turn most difficult screws.

How To Select Wrenches

There are several styles of wrenches. The *open-end* variety (Fig. 2-8) resembles a nut that has one side missing. The openings are on both ends of the handle. *Box-end* wrenches can be identified by the rounded, washer-like nature of the end of their handles.

Combination types combine the features of both. They have open nuts on one end of the handle while the other end resembles a rounded washer. Open-end, box-end and combination wrenches are available both individually or in sets. Each wrench is intended to fit only one size.

The *crescent* wrench is a modification of the open-end type. It is adjustable to fit several different nut sizes. However, when using this type of wrench it is necessary to check periodically to make certain that the jaws fit tightly around the nut.

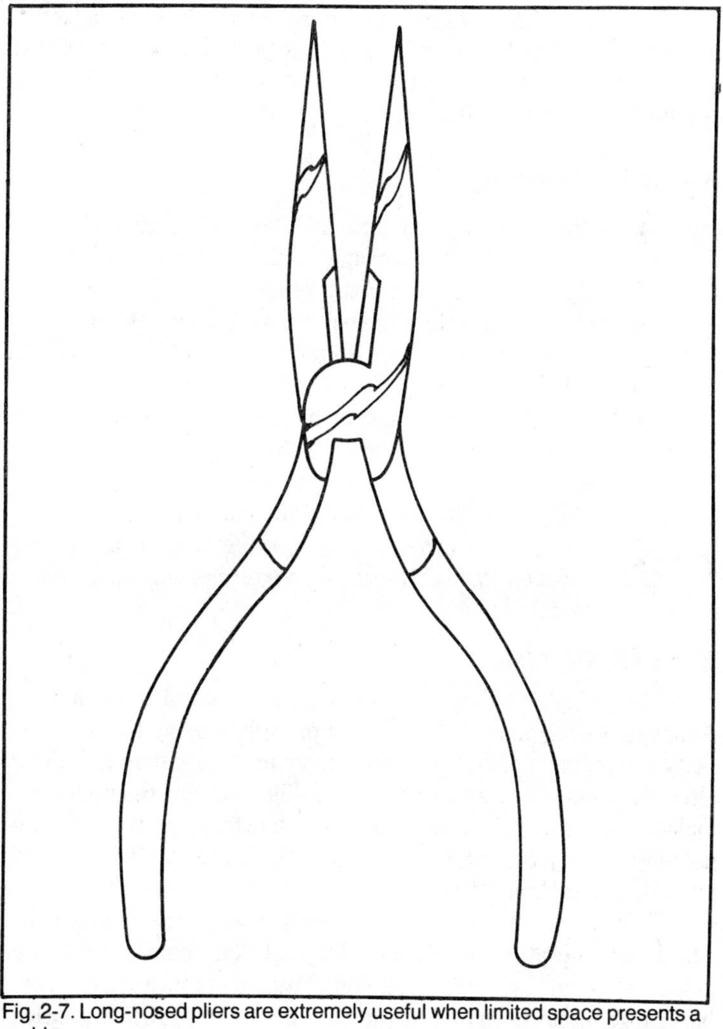

Fig. 2-7. Long-nosed pliers are extremely useful when limited space presents a problem.

Open-end wrenches are particularly convenient for turning nuts in tight places where you can't just spin them on with your fingers. They are also usefui for keeping nuts from turning while you turn bolts to tighten them. The main advantage of the box wrench is that it allows you to exert considerable force without worrying about the wrench slipping.

There are other types of wrenches available. However, they are generally not used in carpentry. *Monkey* and *pipe* wrenches should not be used with nuts and bolts since their jaws are not parallel and they will damage the nuts and bolts.

If you do not have experience in using wrenches, an inexpensive set of combination wrenches is your best bet. Augment them with a crescent wrench to fit nuts that can't be accommodated by the combination wrenches.

How To Use Wrenches

- Choose a wrench that fits the nut snugly. This will prevent the nut's shape, which is usually square or six-sided, from becoming rounded.
- Keep the handle of the wrench at right angles to the bolt. Remember you want to use a wrench for turning the nut—not prying it off the bolt.
- Hold the handle of the wrench firmly near the end. This will keep you in control of the tool and make rotation easier.
- Make sure that as much of the nut as possible fits into the wrench. The chances for the wrench to slip are reduced while at the same time turning the nut becomes easier.

TOOLS FOR DRILLING

The ability to drill holes is extremely important to carpentry. Although drilling holes in wood is not difficult, making them improperly can result in damaged tools, attachments, and materials. The ultimate result is lost time and money. When you use the right tool, making holes is simple. Knowing how to use the tools properly will not only add to your craftsmanship, it will save time and allow you to use fewer attachments.

Two terms you should become familiar with are *bits* and *drills*. These are the attachments for each tool that actually make the holes. Although the terms are sometimes used interchangeably, bits are generally used to make holes in wood. This process is called

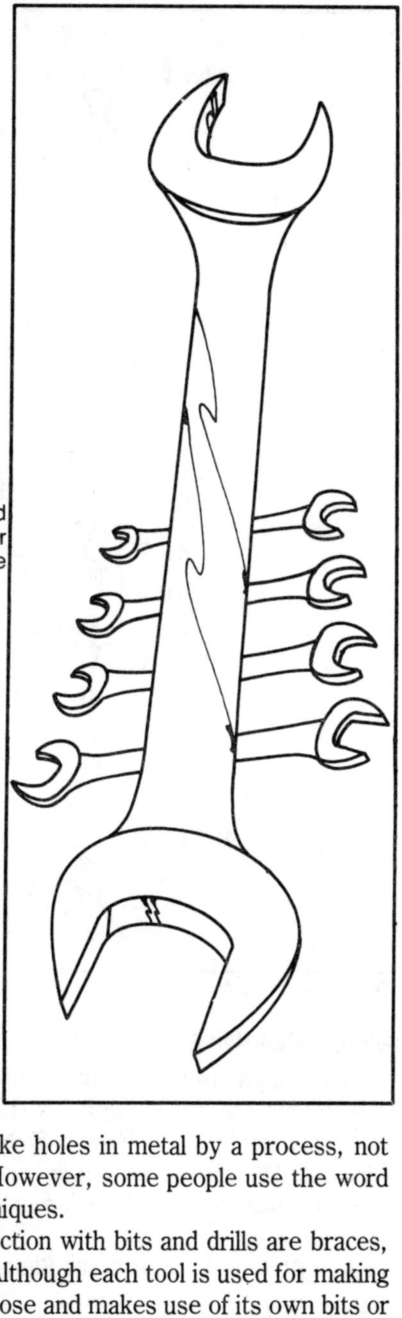

Fig. 2-8. A set of open-ended wrenches. They are convenient for turning nuts or for holding them while bolts are tightened.

boring. Drills are used to make holes in metal by a process, not surprisingly, called *drilling*. However, some people use the word drilling to apply to both techniques.

The tools used in conjunction with bits and drills are braces, hand drills and power drills. Although each tool is used for making holes, each has a special purpose and makes use of its own bits or drills.

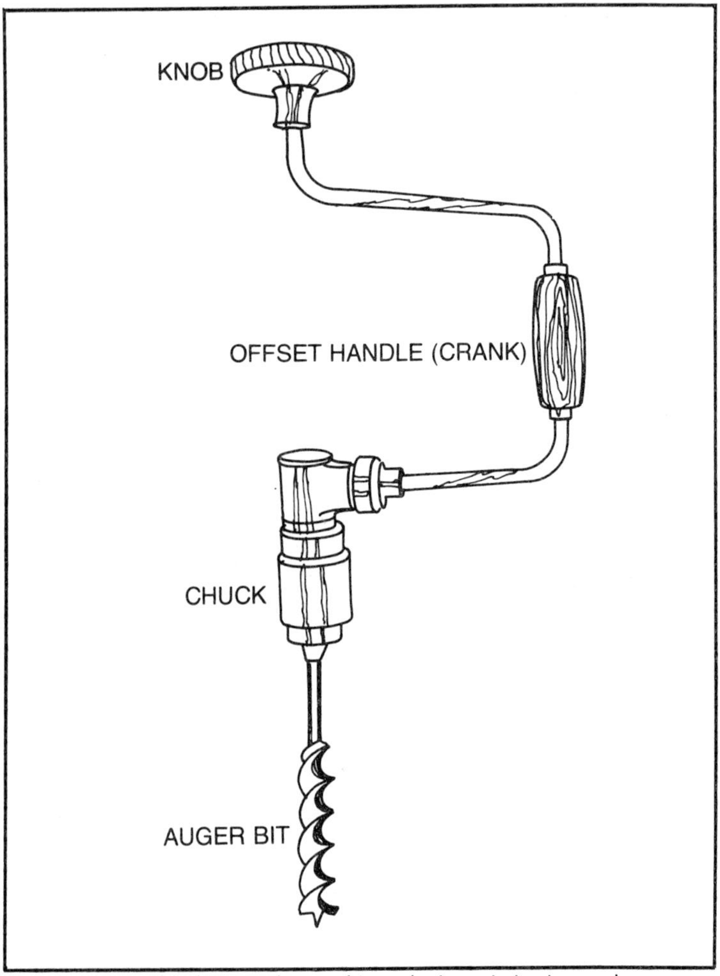

Fig. 2-9. A brace and bit can be used to make large holes in wood.

Boring Large Holes

Braces are used for boring large holes in wood. Its companion attachments are called *auger bits* (Fig. 2-9). Resembling cranks, braces are used to make holes from one quarter-inch in diameter to 2 inches. The appropriate bit is inserted into a part of the brace called a *chuck*, which holds it firmly in place.

How To Place A Bit In A Brace Chuck

- Turn the chuck so that its jaws open just wide enough to allow the bit to be inserted.

- Place the bit in the chuck as far as possible.
- Turn the chuck in the opposite direction to tighten the jaws around the bit.
- Check to make certain that the bit is well centered within the jaws of the chuck. If so, turn the chuck so that it is gripping the bit firmly. If the bit is off center or at an angle, simply turn the chuck to loosen the jaws, re-position the bit, and tighten the jaws once again.

Once the bit is positioned within the brace chuck you are ready to bore a hole.

How To Bore A Hole With A Brace

- Using a pencil, mark the exact spot on the wood where you wish to place the hole.
- Place the point of the bit on this mark. Unless you specifically wish to bore a hole at an angle, be sure that the bit is perpendicular to the wood.

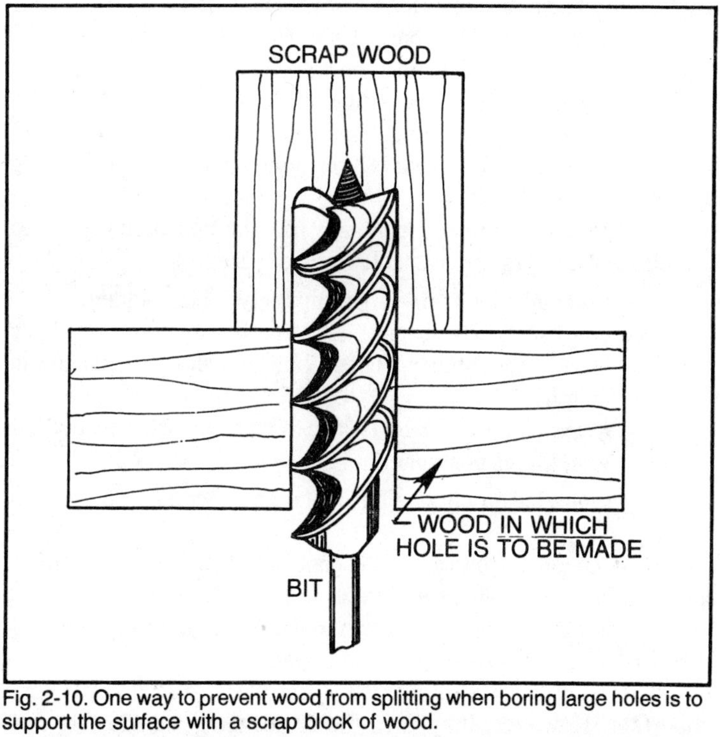

Fig. 2-10. One way to prevent wood from splitting when boring large holes is to support the surface with a scrap block of wood.

- Hold the round knob of the brace with one hand while cranking the handle with the other hand. Continue cranking until the bit is almost ready to pierce the other side of the wood.
- Remove the bit from the wood.
- Place a scrap block of wood behind the hole where the bit would normally pierce it then re-insert the bit in the hole and continue cranking the brace handle until the hole is completed (Fig. 2-10).

Here is an alternative to step five. Mark the spot behind the wood where the point of the bit would emerge. Then place the bit point on this mark and bore from this side of the wood until you meet the hole you have already started (Fig. 2-11). Be sure to use one of the techniques mentioned in step five. Boring straight through the wood will cause it to split.

Because braces are used with a cranking motion, the offset handle swings in an arc. Depending on the tool, this arc will be anywhere from 6 to 14 inches. This presents no difficulty as long as there is enough space available to allow this cranking motion to be made. A modified form of brace allows holes to be bored with just a partial arc being made. The handle is moved only in a short arc and then returned to its original position before the process is repeated until the hole is completed. Step five or its alternative must still be done to prevent wood splitting. This modification is called a *rachet* and is a convenient gadget that is often added to other tools besides braces.

No brace can be used unless it has the appropriate bits to complement it. Bits are available in three sizes.

— *Short dowel* bits are approximately 5 inches long.

— *Long ship* bits are intended for boring through heavy boards and timber. These bits are about 10 inches in length.

— *Regular* bits are 7 to 10 inches long and are intended for general purpose use.

If you purchase only one bit, the regular type is the one you should buy.

Besides length, bits are also constructed differently. A *hollow spiral* variety, also called *single twist*, is intended for deep boring. It is made so that cuttings are thrown to the center hole which allows them to be easily delivered to the surface.

The *double twist* bit is designed for use where accuracy is very important. However, they have the disadvantage of allowing cut-

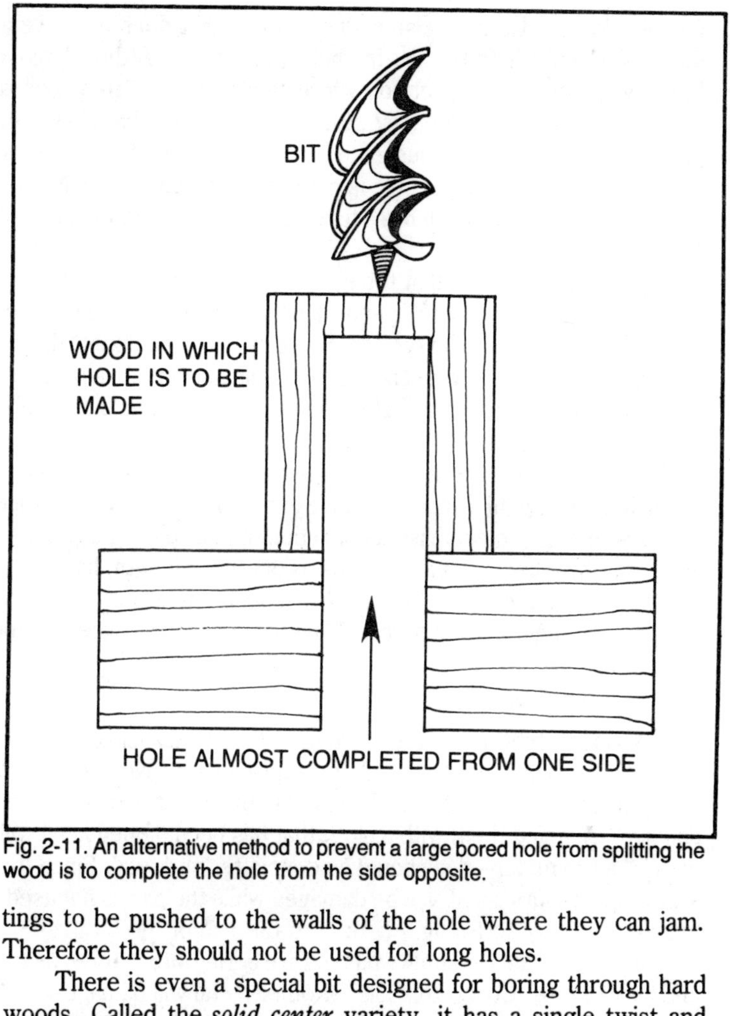

Fig. 2-11. An alternative method to prevent a large bored hole from splitting the wood is to complete the hole from the side opposite.

tings to be pushed to the walls of the hole where they can jam. Therefore they should not be used for long holes.

There is even a special bit designed for boring through hard woods. Called the *solid center* variety, it has a single twist and resembles a screw thread wrapped around a metal rod.

Brace bits also are constructed with different types of heads, each one designed to assist the bit with a specific purpose.

The portion of the head that actually knifes into the wood is called a *cutter*. Some bits have one or two pieces of metal that stick up from the thread. These are called *spurs*.

There is a *double cutter* with a *double spur*, a head that is excellent for general use for making holes that are smooth since it bores and cleans easily. The *single cutter* which has a spur is intended primarily for use when the wood is wet, unseasoned, or contains knots. There is also a single cutter that does not have a spur and is constructed for use in a hole that is rough. Although most bits have a screw thread on top which allows the hole to be bored accurately, there is a *barefoot singlecutter* variety that does not possess one. This type is designed for boring in wood that is wet or pitchy and when a straight hole is not a requirement. Since it does not contain a screw, the bit has a tendency to move with the grain of the wood.

The type of thread that the screw contains affects the type of hole that is made and the speed at which the boring can be accomplished. Single thread types pull rapidly while the double thread variety pulls more slowly. The thread itself can contain a different number of turns. Some will have about six turns per inch, while others will contain up to 18 turns.

With the wide variety of bit types and construction, which is the best one for you? Unless you have experience in the use of brace and bits or want to establish a collection of fine, different purpose bits, your best bet is a quality bit that is of regular length, has a solid center with a single twist, is built with a double spur and cutter and contains a single course thread. This type of bit is excellent for carpentry.

How To Take Care Of Brace Bits

Although bits are made of rugged metal, misusing them or neglect can result in the need for replacement. Keeping them wrapped in folded cloth will help keep the cutting edges from damage. Cleaning them with a cloth that has been dipped in oil will preserve the metal—this should be done after each use. Since the screw tip or spurs could also be damaged while the part is not used, protect them by screwing a cork onto the end of the bit.

It is a good idea to know what you are boring into before you try and make a hole. Concrete, nails, or other metal will damage bits.

DRILLS

Drills are some of the most useful tools in any carpentry workshop. Available in manual or electric models, their original intent was for drilling small holes. While boring holes up to one quarter-inch is still the main purpose of the hand drill, the electric drill has become one of the most versatile of tools.

MANUAL DRILLS

Manual drills enable you to make small holes in wood much more rapidly than through the use of braces. They are available in three types. Push drills, as their name implies, work by pushing their handles toward the wood in which the hole is to be made. Breast and hand drills have a crank that is turned in a circular motion. The only difference between the hand drill and the breast type is that the latter variety has a bar that rests against your chest and allows you to exert greater pressure while the hand drill has a handle.

HOW TO USE HAND DRILLS PROPERLY

- When drilling wood, start by making a small hole with a hammer and nail.
- Place the center of the drill bit in the hole you made with the nail. Unless you wish to make a hole at an angle hold the drill so that the bit is at right angles to the wood.
- Be sure to hold the drill straight. Bending it sideways can result in broken bits.
- Be certain the bit you select is positioned properly in the drill chuck. If it isn't, it will make a hole larger than you want.
- As you are about to complete the hole, slightly reduce the pressure. This is particularly important with metal and some hard woods to prevent the point of the bit from breaking.
- Be sure to keep revolving the drill as you remove the bit from the completed hole. When the drill is removed abruptly the bit often breaks.

Electric Drills

Electric drills are so useful that they are made in more styles and types and by more manufacturers than any other portable power tool. The numbers and variety of accessories almost make this tool a workshop in itself. Drills are available in light, medium and heavy duty classifications.

Light-duty drills generally average about one eighth-horsepower and are convenient for hole drilling and many other simple jobs. However, for frequent use or to fully avail yourself of the wide capabilities of the tool, a medium duty or heavy duty model is recommended.

One of the most commonly used parameters for determining drill capability is the diameter of the chuck. A chuck is the part of the drill that holds the bits and other attachments. Generally, one quarter-inch chucks are used on drills intended for light to medium duty. Three eighth-inch and half-inch varieties are designed for medium to heavy duty usage. The three eighth-inch and half-inch models usually cost more than the more common one quarter-inch variety, but they deliver the extra power required when working with difficult materials and when you are using accessories.

The chuck on a portable electric drill is made with hardened steel jaws to grip the bits and other attachments. Some of these chucks are constructed so that hand pressure is all that is needed to tighten them. Others require the use of *keys*. The key types (Fig. 2-12) are more common and have teeth on the lower edge. The key has teeth that complement those on the chuck. The key is positioned so that when the key is turned the jaws of the chuck are tightened around the bit or attachment.

How To Use A Key-Type Chuck

- Using your hand, rotate the chuck until the jaws open slightly wider than the base of the bit or attachment that you wish to use.
- Insert the base of the bit or attachment as far as it will go between the jaws.
- Make certain that the bit or accessory is centered properly within the jaws. If it isn't it can break and ruin your job.
- Insert the key into one of the holes in the chuck so that its teeth engage those of the chuck.
- Turn the key so that the jaws of the chuck tighten around the attachment. A moderately firm pressure is all that is required.

Electric Drill Features

To considerably reduce the chances for accidental electrical shock when using any electric tool, it should be *grounded*. Basically, what this means is that somewhere there is an electrical circuit that connects with the earth. Usually, grounded electrical tools are identified by three-wired plugs that fit into grounded outlets (Fig. 1-1). The long rounded prong on the plug is part of the grounding circuit.

There is one drawback to the three-wire grounding method. If something happens to the grounding circuit (for example, a loose wire, an improperly connected outlet or the inadvertant use of an

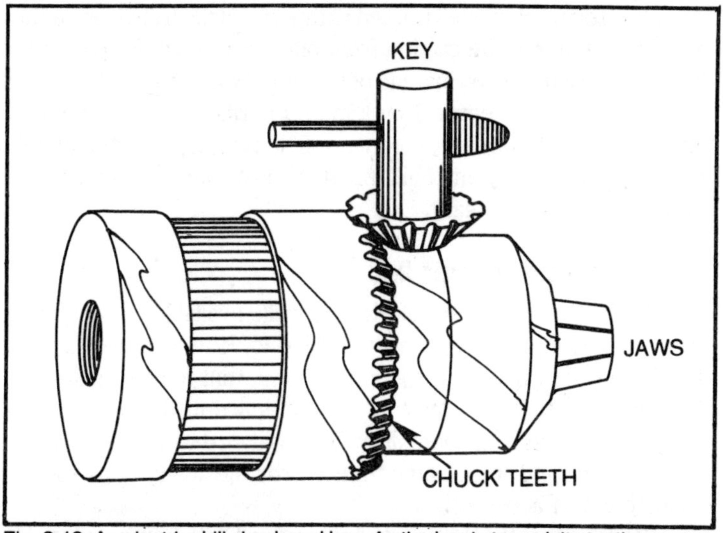

Fig. 2-12. An electric drill chuck and key. As the key is turned, its teeth engage with those on the chuck and the jaws of the chuck are tightened around the chosen bit or accessory.

ungrounded extension cord) the possibility of electrical shock is still present. This is particularly true if you are working in a damp area, your hands are moist or the material you are working with is wet. To avoid this problem, many electrical tools are now manufactured using the *double-insulated* technique. They have two separate and distinct insulating systems. Each insulation is built with protection against mechanical injury. In the event of failure in one of the insulations, the other would provide protection. The double insulation feature replaces the grounding technique in these tools. Therefore the grounded plug is unnecessary. These models have standard two prong plugs. **Caution:** If any electrical tool has a two-pronged plug, be sure that it is clearly labeled as double-insulated. Do not assume that because a tool has a two-pronged plug that it is double-insulated.

One feature offered on some models is the ability to control the speed at which they operate. These models have an electronic speed control that is part of the trigger circuit so that as the trigger is squeezed the speed is increased.

While electric drills may look similar, their construction can differ markedly. For example, every drill has a handle. However, these handles are located in different positions on the tool and they have a variety of shapes. Some heavy-duty types even have an extra assist grip to aid in steadying the tool.

The term "portable" is used with an electric drill because they are light enough to be carried from one place to another. You can bring the tool to the job instead of the other way around. However, they still require electrical outlets and perhaps extension cords. There is one version of the electric drill that is truly portable. It has a rechargeable energy unit built into it and is designed so that special grounding is unnecessary.

How To Select An Electric Drill

With the assortment of electric drills available, selecting the correct one for your needs can be confusing. However, because this tool is so widely used in carpentry you will want to be satisfied with your purchase. If you don't have experience in the use of drills, select a drill that is medium duty. This will enable you to do a wide range of jobs and at the same time affords the ability to use a large assortment of accessories.

Be sure that the drill is rated at least one-fifth horsepower. The higher the horsepower the "stronger" the tool—but the greater its cost. It should be grounded or double-insulated. Choose a drill with either a one-quarter-inch chuck, or preferably a three-eighth inch type. The variable speed control is a useful feature. Although drills with this option cost more, it will probably be worth the additional charge to have this convenience.

Select an electric drill that is comfortable for you to hold. It should feel balanced to you and not be too heavy to operate comfortably. The trigger should be easily accessable to your finger when you hold the tool with one hand. A trigger lock enables the drill to be operated without the trigger being continually depressed. Be sure that you can use it without difficulty.

As long as you are examining the trigger, plug in the tool and operate it. If your finger is uncomfortable your hand will become tired rapidly during heavy drilling jobs. Consider a different model.

While the drill is operating, listen to its sound. Although the sound level will differ with different models, the sound should be free from "strange noises"—a sign of possible gear chatter.

Unplug the unit and hold the housing firmly in one hand while grasping the chuck with the other. There should be little or no play in the bearing fit.

How To Use A Power Drill Properly

Misusing a power drill results in damaged twist drills or damage to the unit itself, as well as affecting the quality of the finished product. Using a power drill properly will save you time lost through carelessness.

Because power drills are lightweight, there is a tendency to use them with only one hand. However, using both hands (when it is not necessary to grip the work) will give more control. Apply a moderate pressure with one hand holding the drill handle. Use the other hand to grip the drill body. This will allow you to check the movement of the tool if you suddenly penetrate the work. It also affords the additional advantage of steadying the tool and preventing wobbling that could break the twist drill and result in an inaccurate hole being made.

Electric Drill Accessories

Despite the large assortment of accessories that can be used with a portable electric drill, its main purpose is drilling holes. The basic part for accomplishing this is the *twist drill.* These have a uniform diameter over the entire length, a smooth shank inserted in

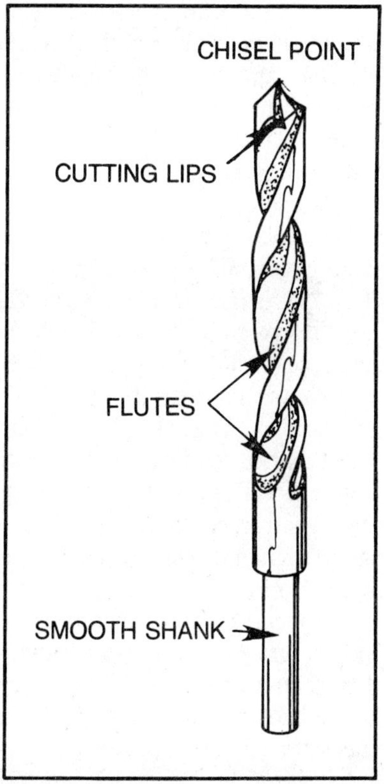

Fig. 2-13. The basic part used with electric drills is the twist drill.

the jaws of the electric drill chuck, spiral flutes that lift chips from the hole being made, two lips for cutting and a chisel point. The construction of a twist drill is illustrated in Fig. 2-13.

Twist drills are made with either carbon tool steel or with high-speed steel. Those constructed with carbon steel are inexpensive and are fine for boring holes in wood. However, the friction of a twist drill turning against wood or metal causes the drill to get hot. Since heat damages these parts, care must be taken so that they do not overheat. Carbon steel twist drills will become dull much more quickly than the high-speed type.

High-speed steel is actually an alloy of tungsten, chromium and vanadium. It derives its name from its ability to retain hardness even at the high temperatures encountered at high speeds. The cutting edges of carbon drills would soften at these speeds.

Most individuals familiar with electric drills prefer high-speed twist drills. Even though they cost more than the carbon steel type, the use of carbon steel drills can actually be more expensive in the long run if much metal drilling is done.

Drill bits come in various lengths and sizes. The greater the assortment you possess, the more flexibility you have. However, a full set is expensive. One-eighth inch and one-quarter inch drills are frequently used items in carpentry. If you do not wish to obtain a full set of bits immediately, start off with these and acquire others as you need them.

In addition to twist drills, there are other bits intended for making holes with power drills. *Spade* bits are generally used to make larger holes than twist drills and are available in sizes from three-eights inch to 1½-inches. They are flat with a sharp point, inexpensive and cut clean holes. Since they are not used as frequently as twist drills, purchase them as the need develops. Figure 2-14 shows a spade bit.

Locator bits are handy for drilling holes in wood that must align with those in hardware. These drills contain a bit inside a metal guide tube. The tube is positioned in the holes in the hardware and the aligned bit bores the hole in the wood. Holes for many purposes can be aligned without the use of this special drill if they are carefully marked prior to drilling.

For cutting large holes in plastics, sheet metal, plywood or thin wood, *hole saws* are convenient, inexpensive attachments. Although they are available in a wide variety of sizes, some of the larger ones are intended only for use in a drill press. Be sure to check the label before using them in a portable drill.

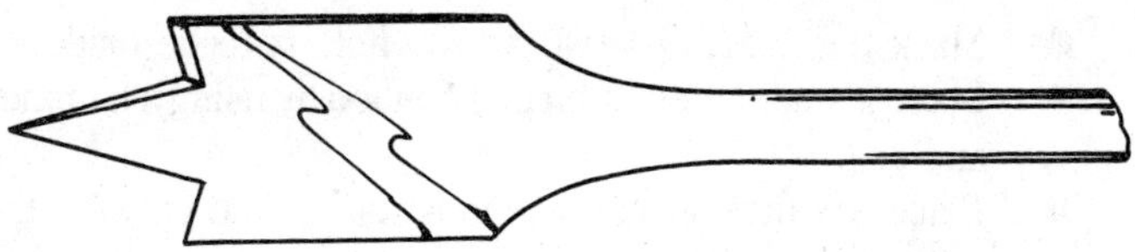

Fig. 2-14. Spade bits are used to make larger holes than you can obtain with standard twist drills.

WHEN A DRILL IS NOT A DRILL

One of the versatile features of the portable electric drill is its ability to be converted into other tools. The power drill can be used as a saw, a sander, grinder, plane, polisher, pump or even garden tools such as a hedge trimmer. If the electric drill can do all this, why is there a need for other power tools?

The answer is that each of these drill attachments is a compromise. For example, the same motor drives each unit rather than one intended for a specific purpose. The outputs are not the same as those of the corresponding power tool and the ability to do work is not as great. As a case in point, a finish sanding attachment will make approximately two thousands orbits in one minute while an actual finishing sander is capable of over four thousand orbits. Although drill attachments won't work as fast or as efficiently as a collection of individual power tools, they are adequate for most carpentry projects. They do afford the important advantage of providing a power workshop at a minimum cost.

How To Select Power Drill Attachments

The design of drill attachments is extremely important. Initially, they were difficult to mount on the drill and often did not fasten securely. Two-handed operation was often a must and safety features were lacking. While these problems have been solved, the time to consider the power attachments you might wish to have is before you purchase a portable drill. This will enable you to determine if the features you will eventually add will fit your drill properly.

Be sure that each attachment you select can be attached to the drill securely. Slots in the body of the drill provide for motor cooling. Be sure that any attachments do not defeat their purpose by blocking these slots. Examine the drill with each choice attached. The entire unit should be balanced and capable of one-handed operation. Make certain that each attachment has guides and adjustments to assure accuracy. Don't forget to check the safety features.

How to Drill a Hole in Wood With an Electric Drill

- Mark the exact spot where the hole is to be made.
- Make a small indentation in the wood by using a hammer and nail.
- Place the drill bit into the indentation and operate the drill very briefly.
- Drills sometimes have a tendency to shift from a hard section of wood to a softer area or they might follow a grain line. Check to make sure the hole is being formed where you intended.
- Hold the drill firmly, using two hands if necessary. Make sure the bit is at right angles to the wood unless you specifically wish the hole to be at an angle.
- Drill the remainder of the hole.

Drilling Tips

The best way to align holes in different pieces of wood is to fasten them together and drill through both pieces at the same time. This can be accomplished by clamping the pieces together, by drilling one pair of matching holes and then using a bolt to hold the pieces in position while the other holes are drilled or by using small finishing nails to join the wood while the holes are made and then carefully removing the nails.

If it is necessary to drill through metal, you will find that a twist drill will tend to move from the spot where the hole is to be made. To avoid this problem, mark the exact spot where you want the hole to be made. Then make an indentation in the metal by striking a large nail (or a *center punch*,) with a hammer. If the indentation that results is slightly bigger than the twist drill point, the drill will stay in the indentation and the hole will be made properly.

If the drill you select has the variable speed feature there are a few general points to remember. The first point is that the smaller the size of the twist drill the faster it must revolve for efficient cutting. The second point concerns the material being drilled. The harder the material is, the slower the speed.

If you must drill through metal, you will find that the twist drill will get very hot. If you allow this condition to continue other than very briefly, damage will result. The use of lubricants will serve as a coolant, increase the rate at which the cutting can be accomplished and make the task easier. Table 2-1 is a handy reference of the correct lubricating agent to use with various materials.

Table 2-1. Materials and lubricating agents to use with them.

Material	Lubricant to Use
hard steel	turpentine
machine steel	soluble oil
wrought iron	lard oil
brass	paraffin (or none)
aluminum	kerosene or turpentine
wood	none
cast iron	none

While the lubricants shown in the table are the ones specifically intended for each material, they might not always be available when you need them. If necessary, common household substitutes can be used. For example, bacon grease melts from the heat generated during drilling. It can be used with wrought and malleable iron and with steel. Light machine oil or kerosene could also be used.

When drilling metal, be sure that it is securely fastened in place. A twist drill can catch the sheet metal as it pentrates the material. This can result in the metal turning along with the drill bit and a dangerous situation that could result in a serious injury.

TOOLS FOR CUTTING

The materials for carpentry craft projects won't always be available in the exact size for your specific job. When materials are too large they have to be cut. Because the ability to cut properly is important to the carpentry craft practitioner, you should carefully consider how to choose and use the saws, chisels and planes that are used for cutting.

SAWS

Saws comprise the broadest category of cutting tools. Because not all materials are constructed in the same manner, saws differ according to the materials they will be used to cut and the way in which this cutting will take place. Like many other tools, saws come in models for both hand and power use.

Handsaws

The most commonly used handsaws are the *crosscut, ripsaw, backsaw, compass saw, coping saw and hacksaw.* The factors that determine the saw type are its shape, the size of the blade and the position and number of teeth it has. Because all saws have teeth, I will cover this feature first.

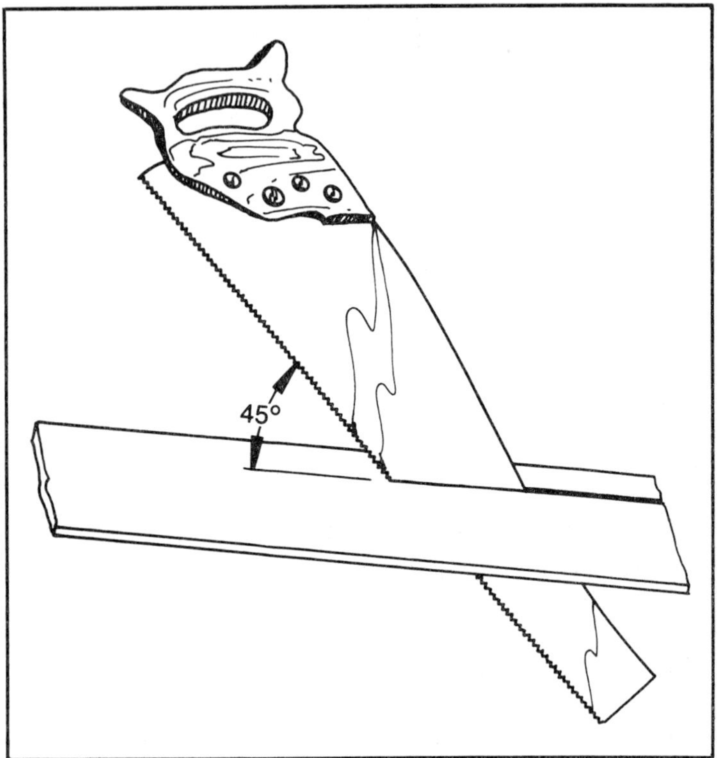

Fig. 2-15. A crosscut saw works best when it makes a 45 degree angle with the wood surface. The tool is designed for cutting across the grain.

One piece of information that appears on a saw is the number of teeth per inch on the saw blade. The term used to describe this number is the *point.* Saw blades with small teeth do their job more slowly and finely than those with larger teeth. Large teeth result in a coarse cut. If the teeth on a saw blade were perfectly aligned with the blade, the blade would not be able to move freely. To overcome this, the teeth are bent slightly outward from the blade to produce a cut that is slightly wider than the blade itself. Saws vary in the way these teeth are bent.

The most frequently used hand saw is the crosscut. This saw gets its name from its intended purpose of sawing across the grain of wood. Actually, the crosscut saw can also be used for cutting plywood, hardboard and other wood items.

The small, chisel-like teeth of the ripsaw allow it to cut rapidly in the direction of the wood grain. Both the ripsaw and the crosscut saw cut the wood as the saw is moved forward. From a cutting standpoint, expending considerable effort to move the saw backward is largely wasted effort.

How To Cut With A Crosscut Saw Or Ripsaw

- Carefully mark a straight line where you intend to cut.
- If you are right-handed, hold the saw handle firmly with that hand. Otherwise use your left hand.
- Crosscut saws work best when the blade is held at a 45 degree angle to the surface (Fig. 2-15). Use a 60 degree angle for a ripsaw (Fig. 2-16).
- Follow the line you made in the first step. Right-handed people generally find it easier to watch the line from the left side of the blade, while those who are left-handed find the reverse to be true.
- Begin cutting by using short strokes.
- Once cutting has begun, use firm, full strokes. Since maximum cutting occurs on the forward motion, relax on the back stroke.
- When you are almost finished with the cut, support the extra wood to prevent it from breaking off.

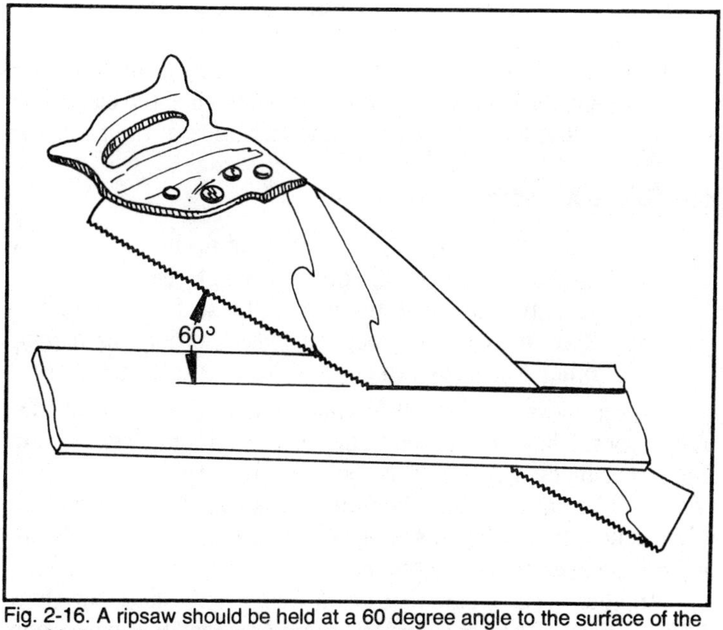

Fig. 2-16. A ripsaw should be held at a 60 degree angle to the surface of the wood being cut. It is used to cut in the direction of the grain.

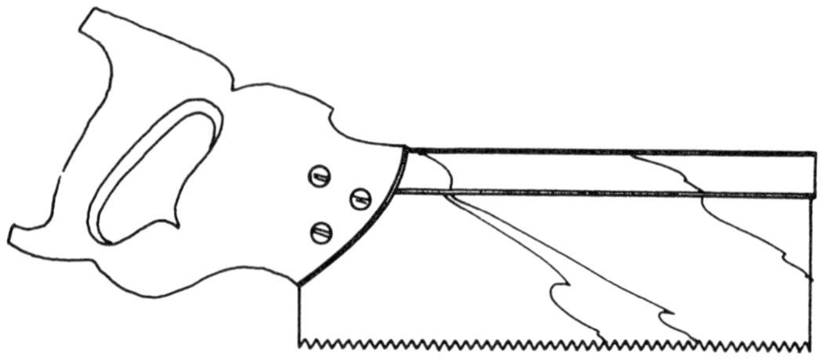

Fig. 2-17. A backsaw allows accurate cutting of straight lines and is recommended for finishing work.

A backsaw (Fig. 2-17) is recommended for finishing work since it allows you to accurately cut straight lines. It has a rectangular blade with a reinforcing strip along its entire length. This strip keeps the blade straight while it is being used and allows a precision cut to be made. It also provides the name of this tool. While the ripsaw and crosscut saw are used at an angle to the item being cut, the backsaw is held parallel to the surface. Since accuracy is usually important with this saw, a *miter box* (an aid in cutting angles and guiding a saw) is often used in conjunction with this tool.

When the job requires cut-outs or curved cuts a compass saw (Fig. 2-18) should be used. A compass saw has a pointed blade that is about one foot long and is about one inch wide at the handle.

How To Use A Compass Saw

- Carefully lay-out the pattern to be made.
- Drill a hole where the first curve is to be made.
- Hold the saw perpendicular to the surface to be cut.
- Start following the pattern. Once cutting has begun, other handsaws can be used for long, straight cuts.

Coping saws (Fig. 2-19) look somewhat different from the other types I have discussed. This tool has a small rectangular frame that holds a thin blade. The blade teeth point toward the handle of the saw and cut when the saw is pulled. Because of its construction, the coping saw is capable of making thin accurate cuts and can be used to saw tight curves.

Another type of saw that has a frame is the hacksaw (Fig. 2-20). Generally used to cut metal, its frame is rigid to reduce blade

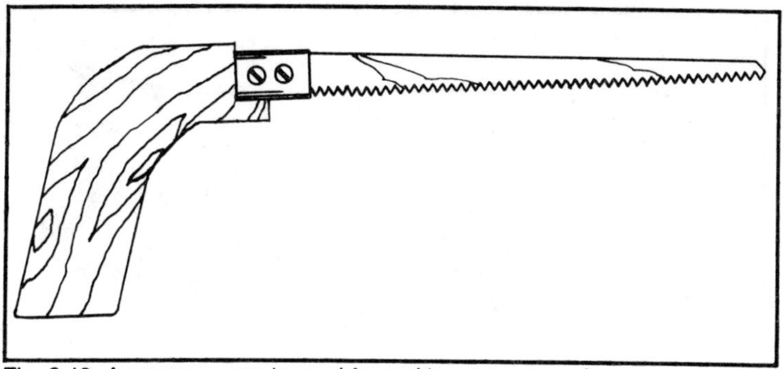

Fig. 2-18. A compass saw is used for making cut-outs and curved cuts.

bending. Different blades can be used with the same frame. Blades having coarse teeth are intended for use on thin materials and those with coarse teeth work well on thick items. The blades are made of high-speed steel. When maximum wear is required, the blade is made from a molybdenum alloy. A tungsten alloy is used in blades designed to cut extremely tough material. The high-speed steel blade is suitable for carpentry craft projects. Hacksaw blades are available with 14, 18, 24, or 32 teeth per inch. While Table 2-2 shows the specific purpose of each of these blades, a blade with 18 teeth per inch is usually best for general use.

How To Use A Hacksaw

- Mark the exact location of the cut on the material.
- Clamp the material so that it can't move while you are cutting.

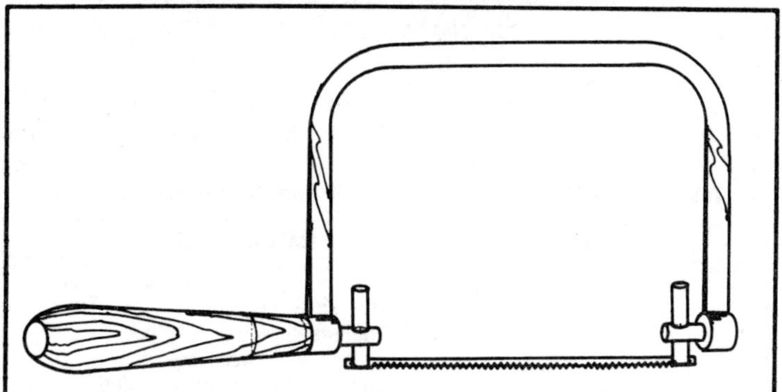

Fig. 2-19. A coping saw is used to make thin, accurate cuts and for sawing tight curves.

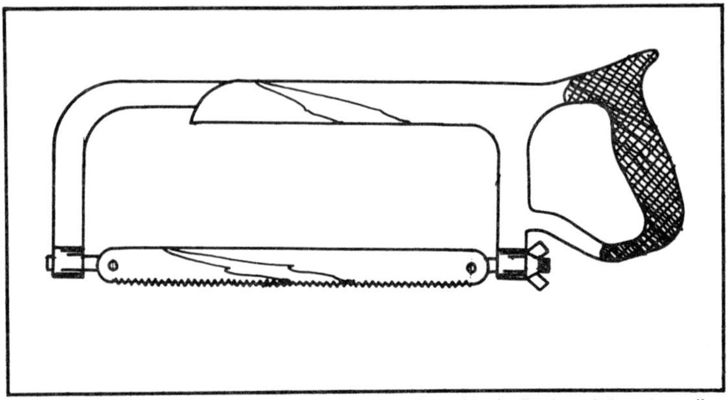

Fig. 2-20. A hacksaw has a stiff frame to reduce blade flexing. It is generally used to cut metal.

- Right-handed individuals should hold the saw with the right hand on the handle with the left hand holding the front corner of the hacksaw frame. If you are left-handed, reverse the procedure.
- Since cutting occurs on the forward stroke, apply moderate pressure with both hands when you move the saw in that direction. Ease up on the pressure during the return stroke.

Power Saws

Portable electric saws have blades that move in either an up-and-down motion or circularly, depending on how the saw is designed to be used. The *sabre saw* (sometimes referred to as a *bayonet saw*) is shown in Figure 2-21. The sabre saw makes use of the reciprocating motion of the blade. It is intended for use in cutting irregular or wavy shapes and straight cuts in some materials. Because it is easy to use and blades are available to cut most

Table 2-2. Hacksaw blade lengths and the materials they can be used on.

TEETH PER INCH	MATERIAL TO BE CUT
14	Aluminum, copper, bronze, brass, iron, soft steel.
18	Tool steel, pipes; best for general purposes use.
24	Sheet metal, tubing; intended for hard materials.
32	Thin materials such as some sheet metal and fine tubing.

materials, the sabre saw is a must for anyone interested in carpentry.

Because of the special designs of different sabre saw blades, it is important to select the proper blade for each job. Blades for use in sawing wood are made from fine tool steel, taper-ground and are razor sharp. They have from six to 12 teeth per inch and they are available in two sizes. The short blade is used for cutting small curves or difficult patterns, while the long variety affords the greater stiffness needed for deep cuts. There are baldes specially designed for cutting metal and even knife blades for cutting paper, leather, and cloth. Table 2-3 illustrates the correct sabre saw blade to use on a variety of materials.

Sabre saws are available in different styles and at a range of prices. Some models have a tilting base so that cuts can be made at an angle. Others have a knob that can be turned to control the position of the blade. This makes intricate cuts much easier.

When selecting a portable electric sabre saw, weight is an important consideration. Although heavy duty models are intended

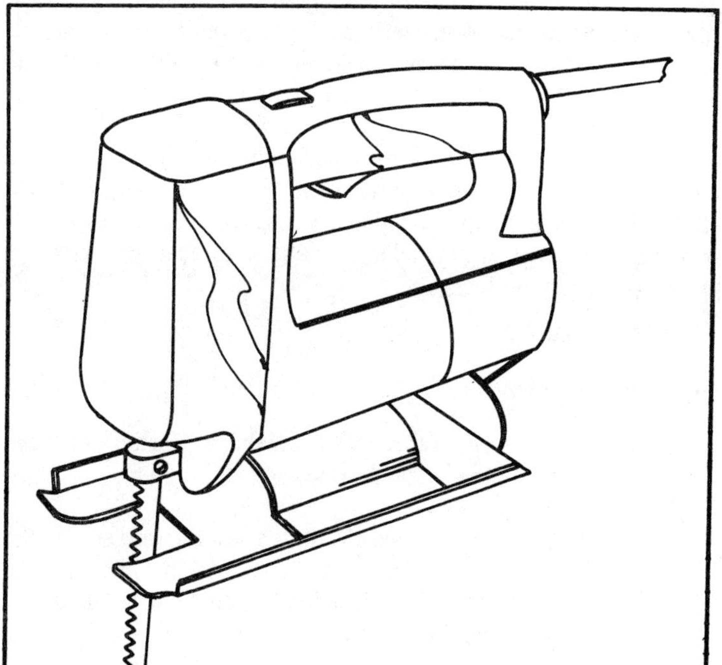

Fig. 2-21. A typical sabre saw.

for long use and for cutting difficult materials, they can be awkward to use in overhead cutting or other positions where the weight of the tool itself must be overcome. Light weight models are fine for carpentry craft projects.

How To Use An Electric Sabre Saw

- Mark the material with the pattern to be made.
- If the surface of the material being cut is flat (such as wood boards) place the front of the saw's shoe plate firmly on the material's surface; do not start the saw until this is done.
- Making sure that the saw's blade does not make contact with the work, start the saw.
- Gently move the blade to the material being cut; the blade will often break if suddenly forced against the material.

Sabre Saw Tips

Because of its practical design, a sabre saw can be used to cut material from the middle rather than the edges as with conventional saws. Since this can be accomplished without the first drilling starter holes, the sabre saw can be used for cutting holes in cabinet tops. An examples would be to cut a hole for a new sink when remodeling the bathroom or for making an opening in paneling so that electrical outlets can be positioned. When an opening is made in the middle of material, the sawing process is referred to as a *plunge cut*. Proper plunge cutting adds to the professional appearance of many carpentry projects.

How To Plunge Cut

- Carefully measure the area to be removed and mark it with a pencil.
- Position the saw over any convenient starting point that is inside the marked area.
- With the saw turned off, tilt the saw forward so that the base is on the surface of the material in which the hole will be made. The blade itself should not be in contact with the surface.
- Start the saw.
- When the blade reaches its full speed, lower the back end of the tool so that the blade penetrates the material.

Table 2-3. Sabre saw blades and the materials they are used on.

MATERIAL	TYPE OF BLADE	NO. TEETH PER INCH	LENGTH
SHEET ALUMINUM		14	
ALUMINUM TUBING	METAL/CUTTING	10, 14	
ANGLE IRON	METAL/CUTTING	14	long
BRASS	METAL/CUTTING	14	
CARDBOARD	KNIFE		
CLOTH	KNIFE		
COPPER	METAL/CUTTING	14	
CORRUGATED BOARD	KNIFE		
FORMICA	WOOD/CUTTING	10	
LEATHER	KNIFE		
LUCITE	WOOD/CUTTING	10	short
PLASTICS	METAL/CUTTING	14	
PLEXIGLASS	WOOD/CUTTING	10	short
RUBBER	KNIFE		
SOFT STEEL	METAL/CUTTING	14	
SHEET STEEL	METAL/CUTTING	14	
THIN STEEL METAL	METAL/CUTTING	24	
METAL TUBING	METAL/CUTTING	14	long

Make certain that the base of the saw rests firmly against the material's surface before continuing.

- Move the sabre saw along the inside of the penciled lines.

How To Cut Square Corners

Sometimes the pattern you want to make contains square corners. Once you have some experience, using a sabre saw for corner cuts becomes easy. To cut corners, use the following steps.

— Follow the inside of the penciled line to the corner.
— Move the saw a short distance back from the corner.
— Moving once again in the forward direction, make a curved cut until you follow the line on the other side of the corner.
— Remove the saw from the material and re-position it so that it is moving toward the corner along the other line.
— Stop cutting when the corner is reached.

Additional Sabre Saw Tips

When a sabre saw is used on certain materials, there might be a tendency for the teeth of the blade to become clogged with small bits of cutting debris. The blade can be cleaned with a small stiff brush. With the power off, of course—remove the plug from the outlet for complete safety. Changing to a coarser blade will reduce the problem.

If it is necessary to interrupt the cutting process to change a blade, be sure to restart from a point other than where you originally began cutting. The new blade might be thicker and the saw will not perform as smoothly if the original, thinner cut is followed.

Care must be taken when thick materials or solid bars are cut. Applying stick wax to the blade before the cut is made makes sawing easier. An added benefit is that the useful life of the tool can often be extended.

Sabre saws don't work properly unless two or more teeth of the blade are in contact with the work. When the teeth of the blade straddle the material, the tool will begin to bounce. If this occurs, change to a blade with finer teeth.

Never force a sabre saw. Electric tools are designed so that the tool should be doing the work—not your muscles. Exert an even force to move the tool forward.

Avoid hurrying. Besides being dangerous, rushing a job results in mistakes that would otherwise be avoided.

How To Select A Sabre Saw

Sabre saws are available with a variety of features and at a broad price range. Less expensive models use a straight reciprocating action and feature a plain, non-tilting base. Because the base is not adjustable, they can only cut at right angles to the work surface. However, they are generally lighter in weight than more expensive types and are easier for beginners to use. In most cases, they are sufficiently powerful for most carpentry projects.

When considering one of these lower priced models, make sure that the tool has a sturdy rotary motor and that the tool is capable of cutting 1½-inch soft wood. It should have a convenient switch on the handle and when operating should generate sufficient blowing action to keep the cutting line clear of debris.

More deluxe models are heavier. Instead of a linear up-and-down motion they are engineered to draw the blade back slightly on the non-cutting downward stroke. This reduces friction and tooth wear on the blade. The blade is moved forward on the cutting upstroke. The resulting orbital action reduces blade wear and permits fast cutting.

For the individual who has some experience in the use of jigsaws, more deluxe models will be attractive. These have a tilting base with a chart marked in degrees to allow sawing at specified angles, blowing action to keep guide lines clear of sawdust and chips and facility for attaching ripping and circle guides. There should be a soft insert in the base that fits snugly against the blade; this permits smooth, splinter free cutting of wood products such as plywood.

Portable Circular Saws

Portable circular saws (Fig. 2-22) are intended for making straight cuts in unwieldy materials. Several times more material can be cut in the same time period by using portable electric circular saws than is possible with the use of hand models. These saws have adjustable controls that allow the depth of the cut to be altered. This is particularly useful when one layer of a material is to be cut but a second layer behind or below it is to remain untouched.

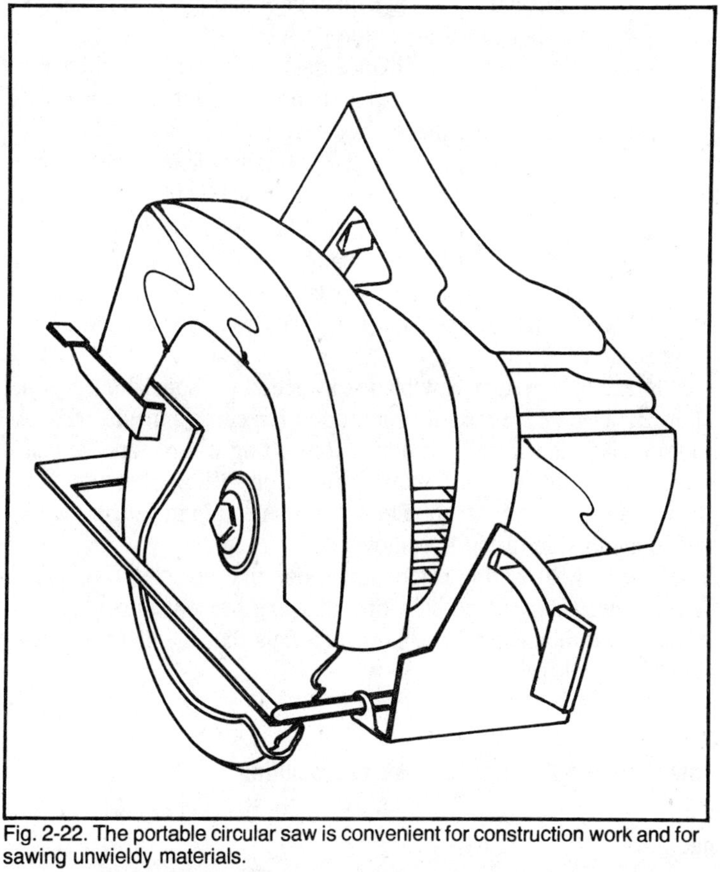

Fig. 2-22. The portable circular saw is convenient for construction work and for sawing unwieldy materials.

The portable circular saw has a circular blade that is turned by a motor through the use of gears. The blade, motor and gears are contained in a housing that has a handle and a trigger switch. The upper portion of the blade is contained in a fixed guard. The bottom part extends through the base of the tool. A moveable guard rotates backward as the wood or other material is cut. This protects the operator from the rotating teeth of the saw blade while at the same time allowing the blade to make contact with the wood. Since saw blades continue to rotate briefly after the tool is switched off, the guard is designed to cover the lower portion of the blade when the tool is removed from the material being cut.

Saws are available with blades from five and one-half inches to more than eight inches. Most of the smaller units will cut a piece of lumber, called a two by four, when the blade is used at right angles to the wood. However, a six and one-half inch blade would be needed to cut the same board at a 45 degree angle. This type of cut is used in a number of construction projects.

There are three types of blades designed for use with circular saws. They are the crosscut, rip and combination blades. For the best results, the correct blade should be used for a specific task. Tooth patterns are shown in Fig. 2-23. It is possible to determine the intended use of a blade by a simple visual inspection. Large, wide teeth mean the blade is designed for rough or deep cutting. Many small teeth indicate that the blade will cut more slowly but will produce cleaner reuslts. Usually, circular saw blades can be classified between these extremes. They are designed to trade speed for precision.

There are circular saw blades intended for special purposes. There is a *miter blade* that is intended for cutting inside trim. A flooring blade is specially designed for cutting where random nails may be present. There is even a blade for cutting plywood. It is engineered to resist the abrasion of glues used in making plywood and to avoid splintering the plywood.

The longest lasting blades are those that are carbide tipped. These have tungsten-carbide tips that are brazed onto a special steel. These blades can cut hard materials that might wear out a standard steel blade.

How To Use a Circular Saw For Crosscutting

Caution: Always make adjustments with the power cord disconnected.

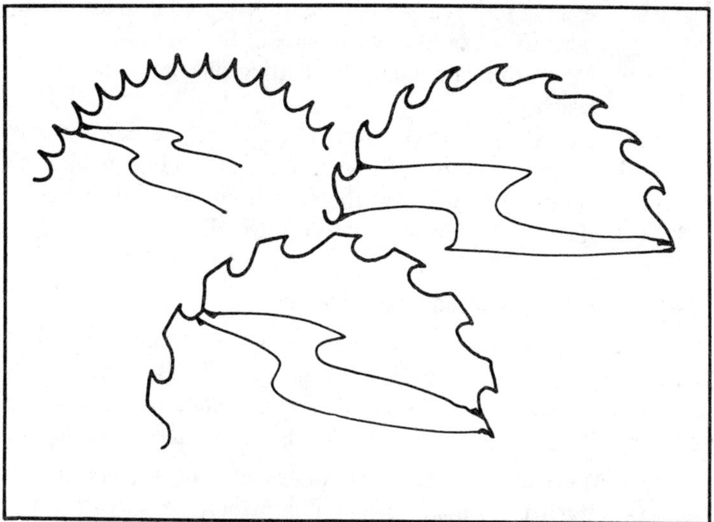

Fig. 2-23. There are three types of circular saw blades for woodworking. The tooth pattern on the left would be found on the crosscut type while the one in the middle is from a ripsaw blade. The teeth on the right would be seen on a combination blade used for rough work.

- Sawing upsupported material is dangerous and usually inaccurate. Be sure the work to be cut is properly supported. Then mark the wood where the cut is to be made.
- Circular saws have a depth of cut adjustment control. Set it so that the valleys of the teeth on the saw blade will clear the material when the cut is made.
- Adjust the angle control so that it is locked at zero.
- Check the guard. It should be in the proper position.
- Connect the power cord. Be certain that it can't get in the way once you have started the cutting procedure.
- With the switch placed in the off position, grasp the handle firmly with one hand while keeping the other hand away from the blade.
- Position the front portion of the base of the saw on the material so that the guide mark on the front plate aligns with the line on the wood to indicate where the cut is to be made.
- While the blade is not touching the materials start the tool and allow the blade to reach full speed.
- Move the saw forward so that the blade engages the wood. Although the line on the wood should be followed

exactly for accurate cutting, it is a good idea for beginners to cut on the waste side of the line.

- Move the saw along the cutting line until the end is reached.
- Release the power switch and move the saw out and away from the wood while the saw blade continues to rotate. Be sure that the saw is not twisted as this is done. Twisting will score the wood.

How To Use A Circular Saw For Ripping

One attachment for circular saws is a ripping guide. This is a piece of metal in the general shape of the letter T. The long end fits into the tool and the other end rides on the edge of the material being cut. If that edge is straight, the rip cut will also be straight.

— After marking the wood where the cut is to be made, attach the ripping guide. For safety, be sure that the power cord is disconnected from the outlet.

— Adjust the ripping guide to the desired width. This is done by aligning a graduation on the guide with a mark on the base of the saw and then locking the guide into position. If the ripping guide with your saw doesn't have graduations, set it so that the guide mark on the front plate of the tool lines up with the line on the wood where the cut is to be made.

— Follow steps 2 through 11 of "How to Use a Circular Saw for Crosscutting", making sure that the material is properly supported.

Circular Saw Tips

If the circular saw stalls while being used, don't release the trigger switch. Instead, move the tool backwards until it regains cutting speed. Then move the saw forward and continue cutting. Excessive stalling is often an indication of a worn blade or an improper setting. Remove the power cord from the outlet and then examine the blade and the saw settings.

Sometimes the cut closes behind the saw blade as the tool is being used. Inserting a wedge, away from the blade, will open the cut and allow sufficient blade clearance.

Always be sure to have enough room for maneuvering. If a long cut will be made, you will have to walk beside the material you are sawing. Be sure that there are no obstacles in your path that you could trip over.

Circular saw blades are sharp and dangerous. Don't rely solely on the guard to prevent an injury. Keep hands and clothing away from the blade.

Keep your eye on the material you are cutting. Besides the possibility of injury, distractions could cause you to make an inaccurate cut.

Speciality Cuts

Portable electric circular saws can be used to make special cuts designed for specific purposes. For example, *miter cuts* are often used to notch rafters in construction work. Miter cuts are easy to make with a circular saw since they are similar to ordinary crosscuts but have an angle of 45 degrees. The angle can be set by making a single adjustment on the angle segment thumbscrew on the saw. *Bevel cuts* are crosscuts with the blade at an angle to the material rather than perpendicular. A combination of the miter cut and the bevel cut is also possible. Figure 2-24 illustrates these cuts.

OTHER POWER SAWS

There are other power saws that are designed for use in the workshop. *Table saws* have a blade that is from seven and one quarter-inches to 12 inches in diameter and located beneath a table surface. Material is pushed through the table saw for cutting. This

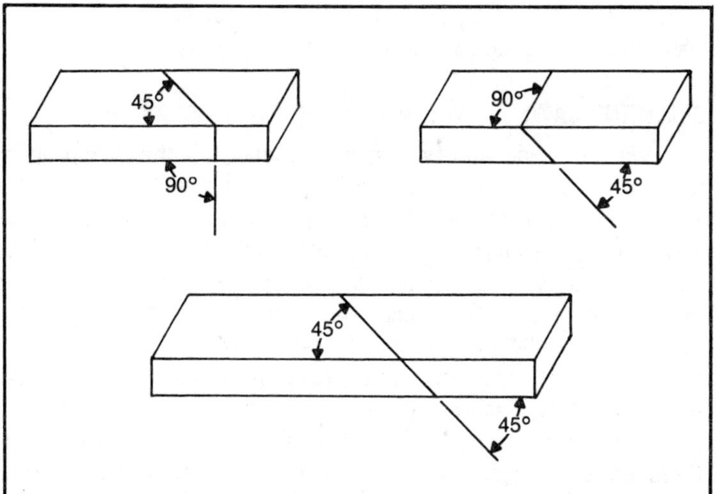

Fig. 2-24. Circular saws can be used to make miter cuts (top left) bevel cuts (top right) and a combination cut (bottom) sometimes referred to as a **combination miter** or **combination bevel**.

type of saw is not portable. A permanent installation in the shop is required. Its major advantage is the high degree of accuracy it provides.

Radial saws are designed to fit above the material. They are safer to operate and more accurate and versatile than table saws. They are not portable and are best positioned along a wall with enough clearance on either side to allow for handling large pieces of lumber. Radial saws can be made to grind, sand, wire brush, drill, rout and shape. While table saws and radial saws accomplish the same tasks, they work differently and each is best suited for specific jobs. Radial saws are preferable for crosscutting. Table saws are more suited for ripping. Because of the expense involved and space requirements for these tools, they are not recommended for the beginner or the individual who wishes to make an occasional carpentry item. Most projects can be accomplished with a sabre saw or portable circular saw.

Another type of power saw is the *reciprocating saw* (Fig. 2-25). This saw resembles an electric knife that might be found in a kitchen. Reciprocating saws are engineered to cut in line with the tool. Because of their construction, they are generally more powerful than comparable sabre saws and they are heavier. Blades are available so that the saw can be used to cut a variety of materials and perform a wide assortment of tasks. Because the blade is aligned with the tool body, this type of saw is particularly useful in sawing one material that is flush against another. An adjustment allows the blade to be positioned vertically or on its side.

SELECTING SAWS

With the wide variety of saws available, selecting the proper tools to meet the needs of carpentry projects can be confusing. Selecting a complete set of hand and power saws will permit quick and accurate sawing for any carpentry project. Unfortunately, it would also be very expensive to fully equip your work area with every type of saw. First gain some familiarity with basic hand models, especially the ripsaw and crosscut types, before considering power saws. Once the basics have been mastered, consider purchasing a good sabre saw. Because of their versatility and ease of use, sabre saws are the most economical power saws. They lend themselves to a wide variety of carpentry projects. If your budget for power tools is limited, a power drill and an electric sabre saw will make excellent choices. If the projects you have in mind do not require hole drilling, the sabre saw should be your first power tool.

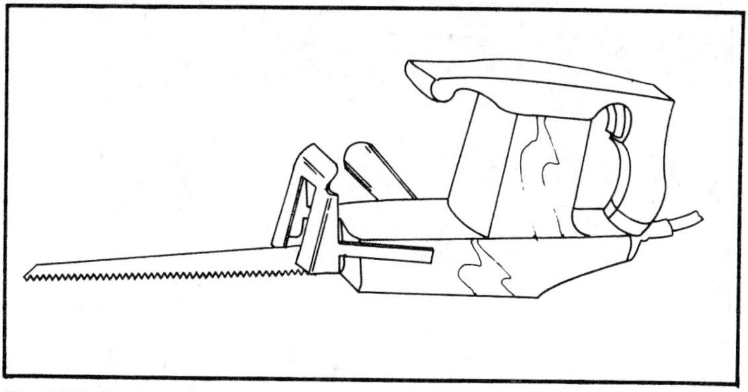

Fig. 2-25. Reciprocating saws are versatile tools used for rough work, scrolling, flush cutting and sawing metal.

CHISELS

Chisels are used when small grooves, notches or cuts must be made. Although they should not be used for a job that another tool can perform better, they are often used when other tools are not available. Chisels are intended to be used to form recesses and *mortises* (square holes into which wooded projections are placed). They can also be used for trimming and paring. A specialized chisel called a *gouge* is used to make rounded grooves.

At one time, chisels had to have thick blades in order to be strong. Alloys are now used that permit the blades to be made thinner so that delicate, as well as heavy-duty procedures can be performed.

How Chisels are Constructed

Chisels are available in a variety of sizes. *Butt chisels* are the shortest type. They have blades that are only about 2½ inches to 3½ inches long and ⅛ to 2 inches wide. The tools themselves are generally under 9 inches in length. Because of their small size, they are ideal for working in limited space.

Pocket chisels are made with the blade approximately 4 to 5 inches long and one-fourth to 2 inches wide. The entire tool is usually 9 to 10 and one-quarter inches long. Their mid-range in size allows them to be used in a wide variety of jobs and makes them one of the most convenient chisels.

The *mill chisel* is the largest type with a blade that is 8 to 10 inches long and 1 to 2 inches wide. The overall length is over 16 inches. Because of their size, mill chisels are generally not used for carpentry craft projects.

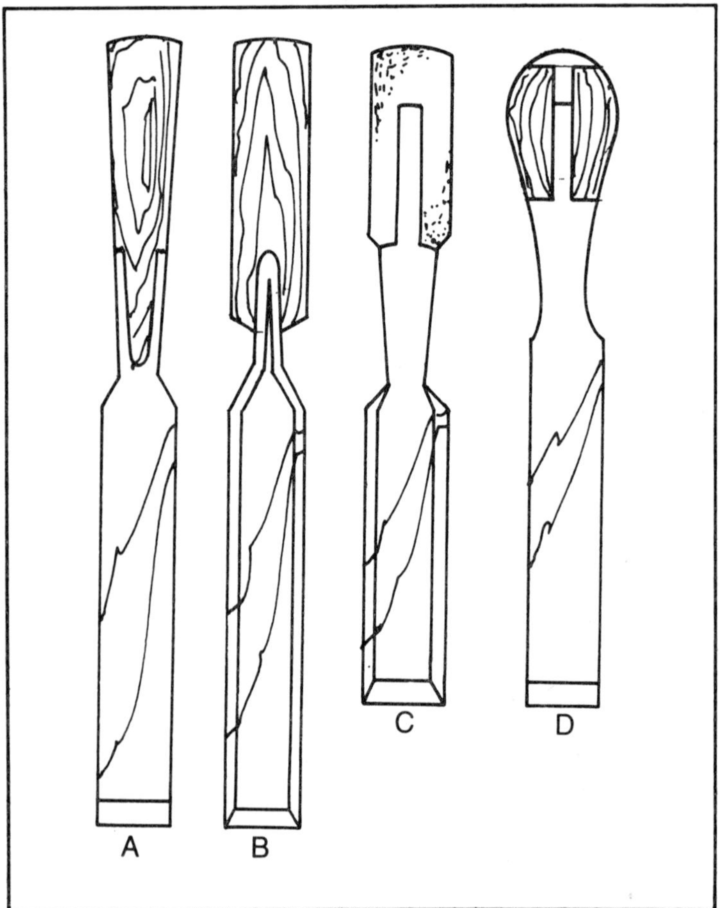

Fig. 2-26. Chisels differ in their construction. A shows a socket chisel. It has a wooden handle fitted to a socket in the blade. B shows a typical tong chisel with a wooden handle. C shows a plastic-handled tong chisel while D details a chisel with a wooden handle and metal core.

Figure 2-26 illustrates the construction of different types of chisels. Chisels differ in composition depending on the task that they will be used for most frequently. Although tang chisels were convenient to use, their wooden handles were frequently split. The modern version of this tool now has a plastic handle that makes it tougher. However, the most durable type has a metal core and a steel cap. The shank of the blade touches the shank of the metal cap so that when the cap is hit by a hammer or mallet, the force is transmitted by metal from the impact point on the cap to the edge of the blade. Although the tradition is to never strike a chisel with anything but a wooden mallet, this version of the tool could be used with any type of hammer.

How to Make a Shallow Recess

- Using a pencil, draw the outline of the recess on the wood surface.
- To prevent the wood from splitting, use a sharp knife to score the lines that run along the grain.
- Place the chisel at one end of the outline. The blade should be positioned across the grain of the wood and the slanted portion of the blade (the *bevel*) should be facing the other end.
- Using moderate force, strike the head of the chisel with a wooden mallet or hammer.
- Repeat the third and fourth steps at the other end of the outline.
- With the bevel of the chisel blade facing in the direction of progress, make a series of cuts parallel to the end cut and about one-eighth inch apart.
- If the recess extends to the edges of the wood, follow the next two steps. Otherwise continue with the previous step.Hold the chisel so that the bevel faces upward and the blade is parallel to the wood surface. Pare inward from the edge Fig. 2-27 to complete the job.
- When the recess is completely enclosed, Hold the chisel so that the bevel of the blade makes contact with the surface of the wood. Move the blade in an uphill direction of the grain to complete the job (Fig. 2-28).

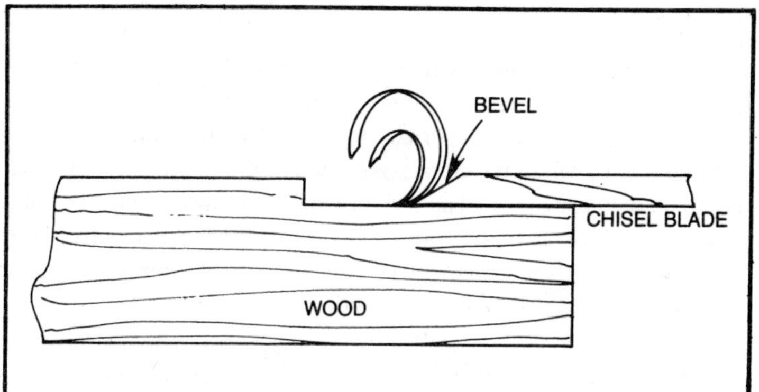

Fig. 2-27. When the recess extends to the edge of the wood, the chisel blade is held parallel to the wood surface with the bevel facing upward.

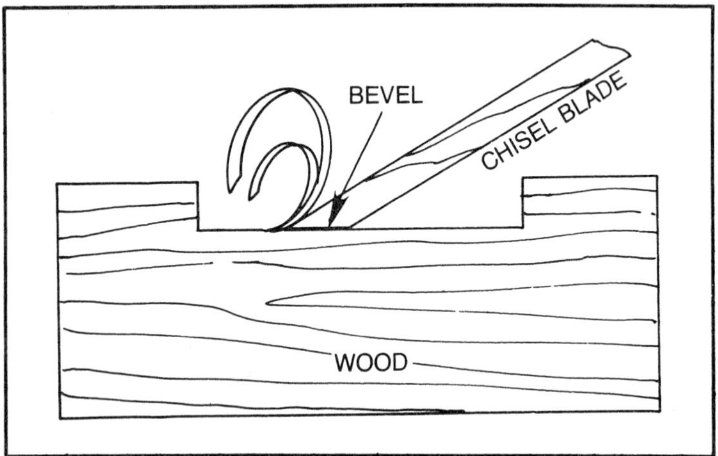

Fig. 2-28. When the recess is in the middle of the wood, hold the chisel so that the bevel of the blade is against the wood and move the tool along the grain.

How to Make a Rectangular Hole (Mortise)

- Using a pencil and straight edge, such as a ruler, mark the outline of the hole.
- Depending on the size of the hole, use a brace and bit or drill to make a series of overlapping holes (Fig. 2-29).
- Using a chisel, pare away the "points" that are left when they are not near the corners of the hole.
- At the corners, use the chisel to cut across the wood grain. This will prevent the wood outside the hole from splitting along the grain. Then, if necessary, use the chisel to pare along the direction of the grain.

Chisel Tips

When using a chisel with a hammer or mallet, be sure that the edge of the chisel is not positioned along the direction of the grain. When the cutting edge is parallel to the grain, the wood will split.

Use care when paring across a wood grain. Best results are obtained by paring inward and slightly upward. Start at one side and work inward toward the center. Then repeat this process from the other side. Remove any material remaining in the center by using horizontal paring cuts from each side.

PLANES

Planes (Fig. 2-30) are designed to slice away excess wood. They resemble a chisel that is placed into a frame. A plane has

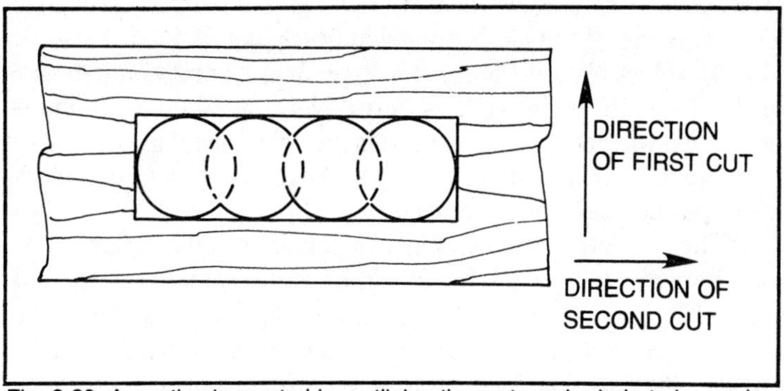

Fig. 2-29. A mortise is created by outlining the rectangular hole to be made, making a series of overlapping holes and then paring the points that remain. When constructing the corners, always cut across the grain first, as shown by the arrows, and finish with the grain.

adjustments to control the depth of its cuts. Besides offering control over the cut it makes, a plane is one of the most convenient and easy tools to use. While care must be taken when using a chisel so that the wood doesn't split, a plane will not split the wood as long as it is used in the direction of the grain (Fig. 2-31). This is because the blade of the plane, called the *plane iron* is positioned within a narrow slit, called the *mouth* of the plane, as the tool is used. Shavings pass through the mouth but this slit isn't wide enough to permit a very heavy slice. If an irregular grain causes a thicker layer to lift, the angle at which the blade is positioned causes the shaving to break near the front edge of the mouth.

Types of Planes

Planes come in a variety of sizes and are designed for specialized purposes such as general woodworking or cabinetry.

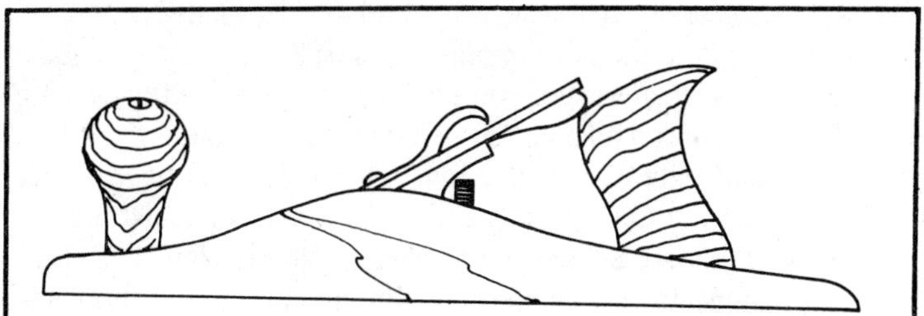

Fig. 2-30. Planes are used to smooth and true wood. They resemble a chisel that has been placed in a frame. The particular tool illustrated here is a jack plane.

However, there are two basic types that are used in carpentry. These are called the *block plane* and *bench planes*.

The block plane is the shortest type. It is generally only about 6 inches long. Its plane angle is normally 12 or 20 degrees and is positioned with its bevel upward. It is the only plane that is intended for cutting along the end of the wood. Because of its size, a block plane can be used with one hand.

There are three types of bench planes, but the major difference between each type is its size. *Fore (jointer) planes* are 18 to 24 inches long and are intended for use when edge-straightening is the objective. *Smooth planes* range from 7 inches to 10 inches long and are designed so that their plane irons are positioned so that the bevel is downward. A cap on the plane iron provides a greater breaking effect on the wood shavings produced as the tool is used. This feature results in minimal roughing to the surface when the plane is used to smooth wood with irregular grains. *Jack planes* are 12 to 15 inches in length and are similar to smooth planes. However, their additional length provides improved edge-straightening qualities. The shorter the length of a plane, the greater its tendency to follow surface irregularities. The general rule to follow is: use a long plane for shaving long surfaces and a short plane for short areas.

How to Use a Plane

- Turn the plane upside down.
- If you are using a bench plane, make sure that the plane iron is mounted so that the bevel is facing dowward. When a block plan is used, the bevel of its blade should face upward.
- Check to insure that the blade protrudes only slightly through the slotted opening, as several thin shavings are preferable to thicker cuts, and that the blade is not on an angle to the bottom of the tool. If it is at an angle, and a bench plane is being used, it can be corrected by pushing the lateral adjustment lever toward the side of the iron that extends the most. If a block plane is being used, release the cam cap and make a manual adjustment.
- If you are using a bench plane, hold the plane at a slight angle to the direction of the cut. Moving in the same direction as the grain, apply most of the force on the front knob at the beginning of the stroke. As the stroke is continued, shift the force to the back handle.

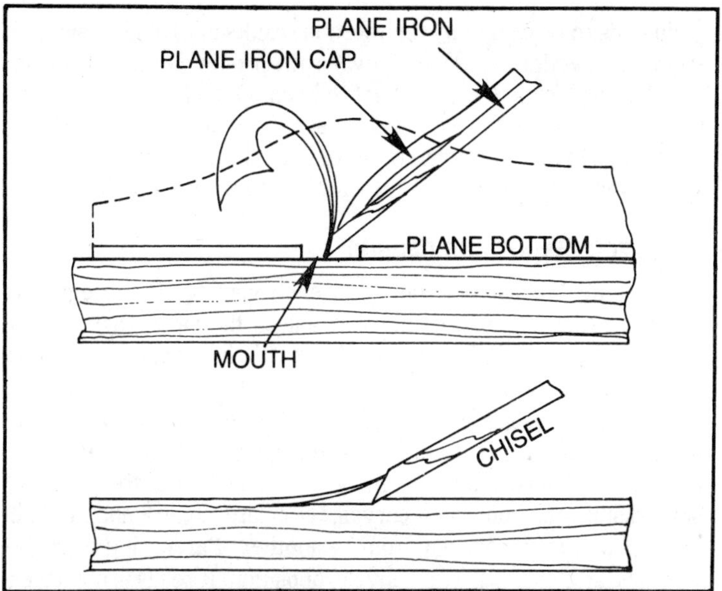

Fig. 2-31. An important advantage of a plane over a chisel for certain types of jobs is that a plane won't split the wood. On the left the shaving is bent by the plane iron (blade). If a thick shaving occurs, it can't fit through the mouth of the plane and it will break. In the same circumstance the chisel (right) would lift a thick layer of wood and splitting would result.

- If you are using a block plane, hold the tool in one hand. If necessary apply pressure to the front knob with your forefinger. When planing the end of the wood, always plane inward from one edge and stop before reaching the other edge. Repeat from the second edge to get a uniform trim. If this isn't done, the board's edge will split. Alternative: clamp a piece of scrap wood to the edge to brace it and cut using a continuous stroke. The scrap wood will split, but the good board will not.

Planing Tips

Most difficulties with planing can actually be prevented before they occur. The first step in preventing trouble is to make sure that the plane is adjusted properly. The blade should be positioned correctly within the frame and a thin shaving should be produced. A few practice strokes on a piece of scrap lumber is a good idea.

Once the plane has been checked, examine the lumber. When planing with the grain, best results are obtained when planing is performed in the "uphill" direction. Going the other way causes the grain ends to be caught by the tool and results in a rougher surface. A quick inspection of the wood reveals the proper direction to plane.

When planing, it is important to keep the tool level and in a line with the surface of the wood unless special effects are desired. For example, allowing the plane to tilt or slant at the start or finish of a stroke will cause slanting of the ends or rounding of the corners.

TOOLS FOR LAYOUT AND GAUGING

Having the correct tools and even gaining some degree of proficiency in their use is fine, but unless the pieces are marked properly they will not fit together. No matter how expertly the tools required to shape and cut the components of a project are used, the results will be disappointing. Before any project can be successfully finished, lines, curves and angles must be measured and drawn accurately. There is an extremely wide selection of tools to measure distances, construct curves, calculate angles and transfer unusual patterns from one item to another. Purchasing each tool would be an extremely expensive proposition. It really is not necessary to do this. Even very elaborate projects can be completed successfully with about a dozen measuring tools. The use of special multipurpose tools reduces this number by about one-half.

Steel Tapes

Steel tape measures are compact and extremely easy to use. Although they are available in lengths of 2 to 100 feet, most are from 6 to 12 feet and retract into a small container. Major manufacturers even design the base of this container to be useful in measuring. The base is exactly 2 inches long. Steel tape rules can be used to measure the distance between items, for example, between the posts of a door frame or to measure an object itself such as the width of a door. The first type or measurement, where the entire ruler is positioned between the end points is called an *inside measurement.* The second type of measurement, where the container portion of the rule would not be used for the measurement itself, is referred to as an *outside measurement.*

In order to make measurements easier, some flexible steel tape rules have a hook loosely rivetted to the end. So that both inside and outside measurements will be accurate, this hook is designed to move a distance equal to its own thickness. Although some individuals mistakenly believe that the looseness of the hook is a manufacturing defect and attempt to tighten it, doing so will introduce an error into measurements.

Quality steel tape rules generally have a locking feature. Once the rule is extended to make a measurement, they will remain extended until a button or switch is pressed to permit the tape to retract into its container. This allows both hands to be free to mark the material. In fact, the average steel rule can be extended unsupported for about 3 feet. This makes it even more convenient to use.

Folding Wooden Rules

Folding rules are made from wooden sections that are from 6 to 8 inches long. These sections are hinged together so that when opened they are capable of measuring up to 12 feet. Its rigidity allows it to be used in many cases where the flexibility of a steel tape would be a disadvantage. Folding rules are extremely accurate and have sliding extensions to permit precise inside measurements. They can also be used to measure along angular shapes or around corners.

Squares

Squares are useful tools for measuring angles and bevels. *Try squares* resemble half a picture frame. They are used to measure right angles and to help you determine if the wood you are working with is square. They are available with blades that range from 3 to 15 inches in length. For convenience, one edge of the tool is graduated for use as a ruler.

How To Use A Try Square

- Place the handle so that it rests firmly against one surface of the wood.
- Slide the edge of the blade into contact with the edge of the wood.
- Check to determine whether light can pass between the blade edge and the wood. If it can, the edge is untrue and should be planed or sanded and then tested again. Continue planing or sanding until light is no longer visable when the check is made.

Combined try and mitre squares resemble the ordinary try square, but have a bevel on the handle so that 45 degree angles can be determined. However, they can only be used to make lines that are only a few inches long.

The *combination square* is one of the most useful tools you can own. If you are going to acquire a limited number of measuring tools, this should be among the first since it can be used to perform several functions. The handle is constructed with an angle of 90 degrees on one edge and 45 degrees on the other so that laying out these angles is easy. A clamp control allows the handle to be fixed at any point along the blade, reversed, or removed completely. Most quality combination squares also contain at least one *level*. A level is a tubular shaped container filled with liquid and containing an air bubble. When the bubble is in the exact center of the tube, the material is level. This type of square also has a convenient *scribe* that is used for marking the wood.

When the blade of the try square is fixed into position, you can use the tool as a convenient *marking gauge* simply by holding a pencil against the tip of the blade. Moving the handle along the blade and then placing the outside of the handle against the wood allows the combination square to be used as a *depth gauge*. Even the handle can be used separately as a level. With the handle in place and the outside of the blade positioned against a vertical surface, the combination square can be used to determine whether the surface is *plumb*, that is, whether it is truly vertical (Fig. 2-32 shows several uses of this tool).

Framing squares resemble ordinary try squares, but they do not have wooden handles. They have sturdier blades and they can be used to rule longer lines. Both sides of both blades are marked with useful scales and charts. Its purpose is to provide a quick reference when carpentry calculations must be made or accurate measurements taken.

BEVELS

Bevels are tools that are usually not used alone. They are generally used in conjunction with a square. They are most often called upon to transfer one angle from one piece of wood to another. This tool isn't often required for basic carpentry projects and you should add it to your collection of tools only as it is needed.

LEVELS AND PLUMBS

Whenever a repair is made on a house or a construction project is built, all horizontal surfaces must be truly horizontal. Similarly, vertical surfaces must always be absolutely vertical. You can make an accurate determination of level and plumb situations even with inexpensive tools. Available as part of a combination square (see above) or as a separate device, levels play an important role in many projects. Levels are often sold in combinations. Each level in the combination could be fixed in a different direction, allowing several

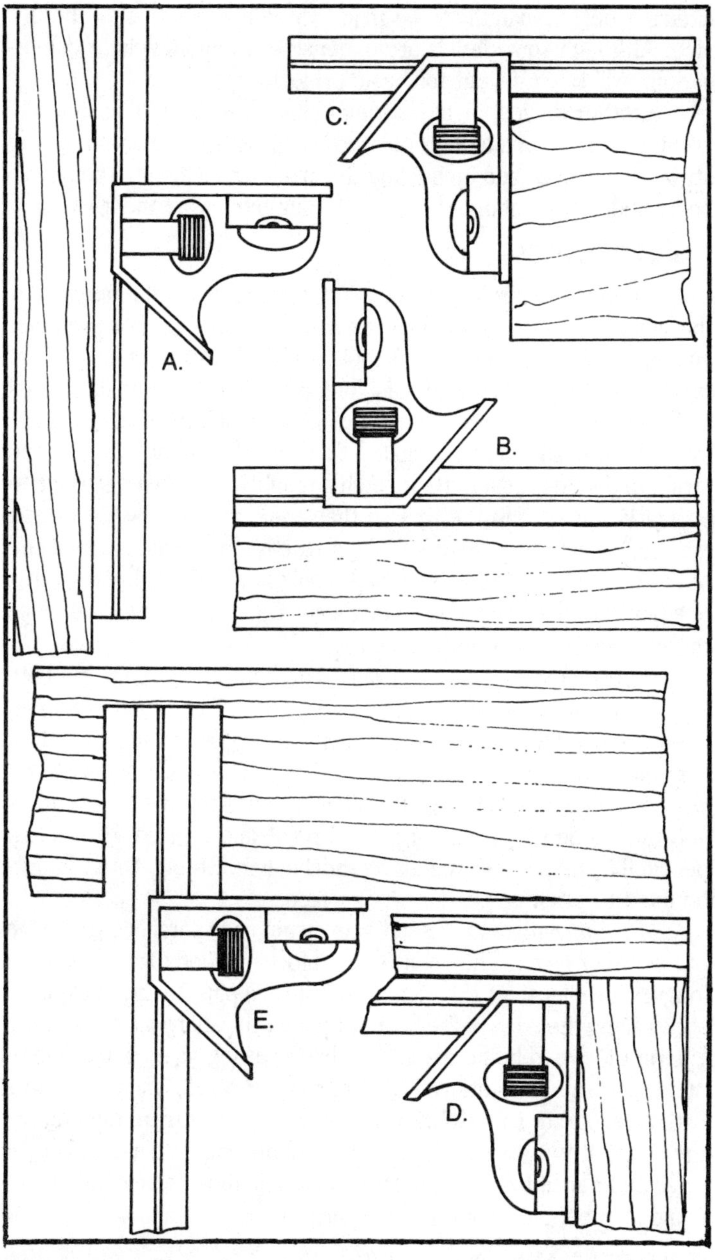

Fig. 2-32. The different uses of the combination square. (A) to determine if a plumb condition exists (B) to measure for level conditions (C) to check whether the outside of something is square, (D) to check whether the inside measurement is square and (E) as a depth gauge.

different determinations to be made without repositioning the device. Although some levels are expensive, simple levels or combinations will be sufficient for basic projects.

Another device for measuring is the *plumb bob.* This is really nothing more than a heavy, pointed weight suspended from a long string. They are used by holding the string and allowing the bob to be positioned just above the ground. Plumb bobs are inexpensive.

MARKING DEVICES

A common marking device used in carpentry is the pencil, but this is not the only item that can be used for marking. For example, an inexpensive and convenient tool for marking straight lines is the *chalk line.* This tool consists of a long cord that is wound on a spool and enclosed in a container that you must fill with chalk. Do not overfill the chalk box with chalk. To make a straight line, pull the string to the desired length, stretch it taughtly across the surface to be marked and snap the line so that a line of chalk is left on the material. While it sometimes helps to have an assistant hold the string taughtly while it is snapped, chalk lines have a hook so that one person can do the job. So that a clear line will be marked, snap the string gently and only once.

Wing dividers resemble ordinary drawing compasses, but they have a special screw that allows them to be locked into position. They are used the same way a drawing compass would be used for scribing circles and dividing arcs and lines. There are two styles of wing dividers. One has two steel divider points for greater accuracy and the second type has one steel point and a place for a sharp pencil. The pencil type makes a readable line. Steel points can only be used to prick or scratch the surface.

Marking gauges are used to construct lines parallel to the edge of a board or panel. They consist of a block of wood through which a longer rule slides. At the end of the rule is a pin. As the thick block slides along the edge of the board or panel, the sharp pin scratches a line on the wood being marked. While marking gauges are convenient, the same results can be obtained by careful measurements with an accurate rule. Mark the desired distance from the edge of the board at two somewhat separated spots and then use a straight edge and pencil to mark the line. A combination square and pencil could also be used for the same purpose.

TOOLS FOR CLAMPING

Clamps are a versatile and important part of any carpentry craft workshop. While they are generally used to hold pieces of wood

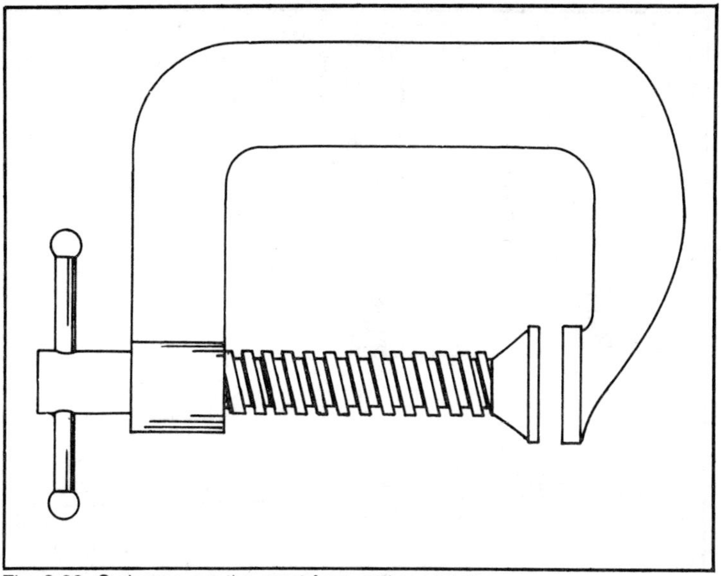

Fig. 2-33. C-clamps are the most frequently used clamp.

under a uniform pressure while the wood is being glued together, they can also serve as a helpful extra pair of hands under certain circumstances. For example, clamps can be used to fasten wooden items to the workbench, freeing your hands to work on the project.

There are various types of clamps. Each type is intended for certain tasks. *C-clamps* (Fig. 2-33) are designed to hold items together or for temporarily fastening items to a workbench. They have jaw widths that vary from 3 to 16 inches and derive their name from their shape. They consist of a frame in the shape of the letter C and have a screw that is used to adjust the holding pressure. *Hand screws* (Fig. 2-34) consist of two pieces of wood that are joined by two adjustable screws. Since both the angle and the distance between the wooden jaws are adjustable, this device is particularly useful for clamping irregularly shaped objects that have flat sides. *Vices* are another form of clamp. They are generally used to hold objects to the workbench. Generally, vices with large jaws are very convenient since they tend to be sturdier and the large jaw area will distribute pressure more evenly over the object being clamped. If you place a vice on the workbench, be sure that the top of its jaws are flush with the top of the bench. *Spring clamps* resemble large clothespins and are used in somewhat the same manner (Fig. 2-35). Normally, the jaws are closed in a clamping position but squeezing the clamp opens them.

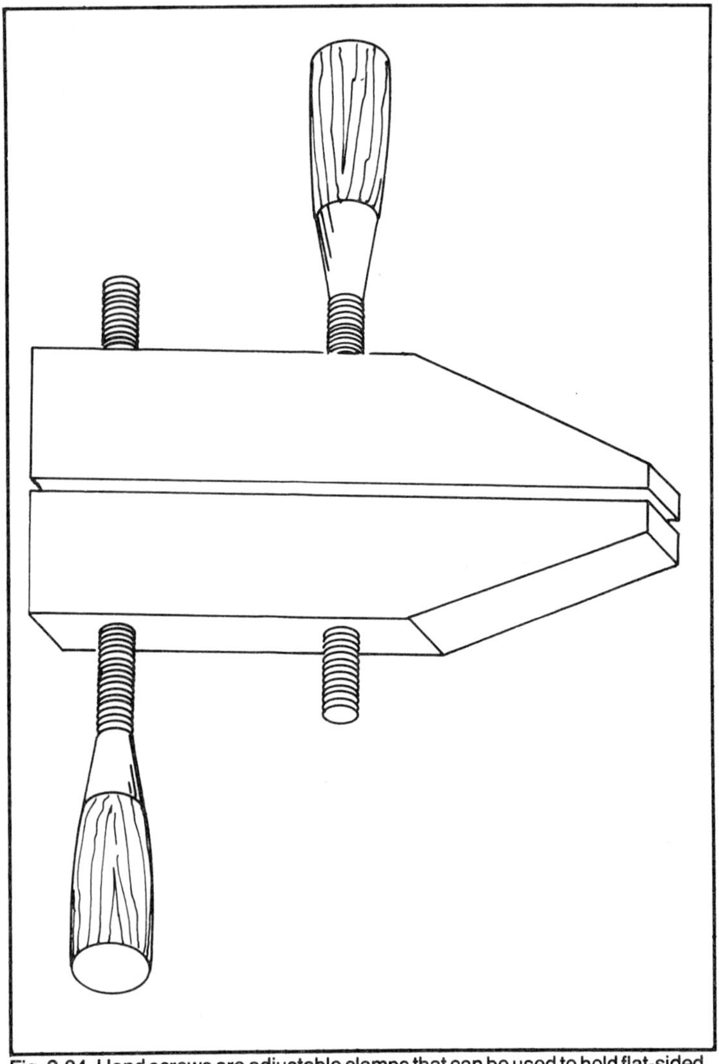

Fig. 2-34. Hand screws are adjustable clamps that can be used to hold flat-sided irregular work. Turning the adjustment screws allows this clamp to be used with angular work. Since they are wooden, they will not damage the surface of the clamped material when they are used properly.

Clamp Priorities

With the wide variety of clamps available, obtaining the correct one for the job at hand can be confusing. However, there are certain guidelines that can be followed. Because of their versatility, C-clamps are the ones you should obtain first and then buy the others as necessary. Some jobs can be accomplished without a special clamp if a little ingenuity is used. For example, a length of rope that can be twisted with a stick, much the way a tourniquet would be used, could do an admirable job of holding items together.

Clamping Hints

The surface of the items clamped together can be damaged by excessive pressure or rough clamp jaws. To reduce the chances of this happening, carpenters frequently place a piece of scrap wood between the jaws of a metal clamp and the good wood surface. In this way, any damage that results will be to the scrap wood and not the material used in the project. This precaution is not necessary with adjustable clamps. Since they are made of wood, using them properly will not result in a marred surface.

HOW TO SAVE MONEY WHEN BUYING TOOLS

Because there are so many do-it-yourself enthusiasts, the availability of quality tools is quite extensive. Hardware stores, department stores, lumber yards, and discount stores are just a few of the many outlets for carpentry tools. Not only do the many stores selling tools offer great convenience to those interested in carpentry but, this competition also provides the potential for saving money on tools as well. The following list describes the ways you can save:

—Browse! By visiting a number of different stores you can compare the costs of tools as well as their availability at each outlet. Examining the assortment available will tell you whether the store deals primarily with brand names or lesser known manufacturers.

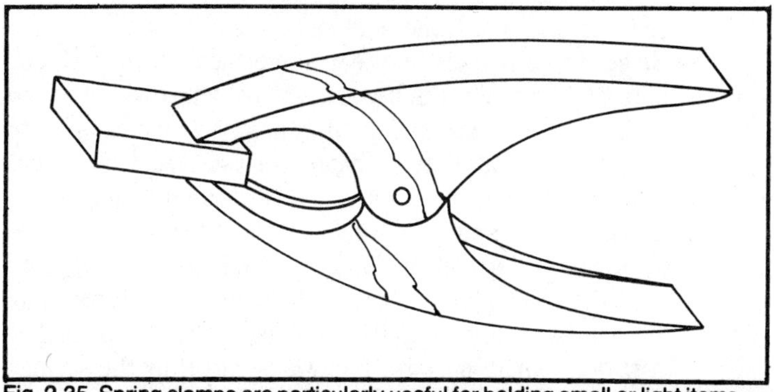

Fig. 2-35. Spring clamps are particularly useful for holding small or light items.

—Check the attitude of the store toward the customer. Do they guarantee your satisfaction and allow returns or exchanges if the customer, not the store, feels it is necessary? If the store specializes in house brands, do they have a written guarantee?

—Does the store have sales personnel familiar with the items they are selling or just individuals who pull items from a shelf or supply room? Because many items cost approximately the same amount no matter what outlet features them, it can be a time-saver to have products demonstrated. This can also save you money later by preventing mistakes that would be made if you misuse the equipment. Remember, mistakes could result in ruined tools or materials.

—Compare prices. Some stores have generally higher prices than others. You should decide if it is wise to purchase a product at a "name" store if the identical item is available from a different establishment at a lower price. In carpentry, it's the way the project turns out that is important.

—Don't be afraid to do comparison shopping. You might even find a tool with better features than the one you were considering.

Look for sales on the particular tools you need. Although you might need a particular item for a project you are working on at the moment, examine newspaper, television and other advertisements for items you will need later.

—Carefully consider the tools you purchase. Do they have expensive gimmicks that you know you will never use? Or perhaps they have a feature you feel could be important on your next project and the additional cost of the attachment or feature is worth the additional price.

—Be sure to buy tools that feel comfortable for you. Heavy duty tools are fine. But if you are not strong enough to use them or if they are designed in a way that makes them awkward for you, they probably won't get much use. Consider a different model even if it isn't rated as highly.

—Consider multi-purpose tools. While tools that are designed to perform many different functions generally don't have the combined features of all the individual tools, they are usually good enough for most jobs. The slightly higher cost of the multi-purpose tool will be offset by the savings involved in not purchasing each tool separately.

—Ask questions! Don't be afraid to show a lack of knowledge about a product. If you don't know how a tool works or how to take full advantage of its features, you will not be getting your money's worth. If store personnel can't, or won't demonstrate the tool for you, go to another store that sells the item. Your money is as good as that of the most experienced craftsman. If the store doesn't recognize that fact and won't answer your questions properly, go to a store that will let you get the most from your purchase. Otherwise, you could buy tools or accessories you don't need.

—Consider whether purchasing the tool is absolutely necessary. If it is an item that you will only use once or twice, find out whether you can borrow or rent the tool. While borrowing tools that you will frequently use is not recommended unless it's absolutely necessary, don't purchase an expensive tool that will remain idle. Also, consider whether an alternative to the tool is available. For example, if the project directions call for a steel rule to be used for measurements, consider whether the same result could be obtained by using the folding rule you do have. Don't be afraid to improvise your own tool. They can sometimes do a specific job better than a general-purpose item. Also, some power tools can be rented. Renting not only provides access to an infrequently used item, but it affords you the opportunity to actually try tools before you consider purchasing them.

WHAT YOU SHOULD KNOW ABOUT POWER TOOLS

You already know that power tools can save time and in many cases they are more versatile than conventional hand tools. But what are the factors that determine whether a particular power tool is capable of performing a specific job? The answer is that each tool has a set of ratings associated with it. The ability to understand these figures and terms allows you to make a sensible determination regarding a tool's performance.

The amount of electrical current that the tool uses is measured in *amperes*. The abbreviation for amperes is *amps*. It might seem that the greater this figure, the more powerful the motor within the tool housing. However, this figure depends upon the load placed on the tool, that is, upon the need of the tool to overcome such things as the friction of the bearings or gears, the air resistance of any fans within the tool and the demands of the actual job for which the tool is being used. The demand for electrical current of a tool increases as the load increases. The greater the load, the greater the need of the tool for current. Because the amount of current actually required by the tool will vary, the rating shown on the nameplate really can't be used by itself as a complete indication of the capacity of a tool.

Voltage refers to the number of units of electrical pressure, called *volts*. The standard voltage for household usage is 110 to 120.

Because the current rating of a tool is not a complete indicator of the tool's capability, manufacturer's usually list the *power rating* of the tool, measured in *watts*. To get this number if it isn't shown, simply multiply the number of volts shown by the number of amperes. If the number of watts is given, but not the current, divide the indicated watts by the voltage to obtain the number of amps. Even the power rating of a tool can be misleading since it depends on the current going into the tool. This current can vary and therefore the actual power will vary.

Many tools have an indication of horsepower, abbreviated hp, that is a rating of the tool's ability to perform work. The higher the horsepower rating, the greater this ability.

To explain how much 1 hp is, imagine a weight of 33,000 pounds being moved 1 foot in the course of 1 minute. A strong man would probably be capable of a sustained effort of at most about one sixth-hp. A horse can only produce about two thirds-hp on a steady basis.

Some tools have a *torque* rating. This is a measure of "twisting" power. This twisting power is really the result of a force being applied to the working end of the tool through a distance from the tool. For example, imagine a shaft that has a one foot long extension to its end. At the end of the extension, picture a 10 pound weight (see Fig. 2-36). The result is 10 foot-pounds of torque on the shaft. Some people have a tendency to associate torque with movement. Actually torque is independent of movement. Referring to the example again, if you were to hold the shaft so that it couldn't turn, the weight would still result in a torque of 10 foot-pounds on the shaft.

Another measurement often found on the nameplate of a tool is the number of *revolutions per minute*, abbreviated *rpm*, which tells how fast the working end of a tool is rotating. There is a mathematical formula that relates torque, speed and horsepower.

Often, the term *cycles*, or *Hertz (Hz)* appears with the other ratings. The *alternating current* (AC) generally used in this country

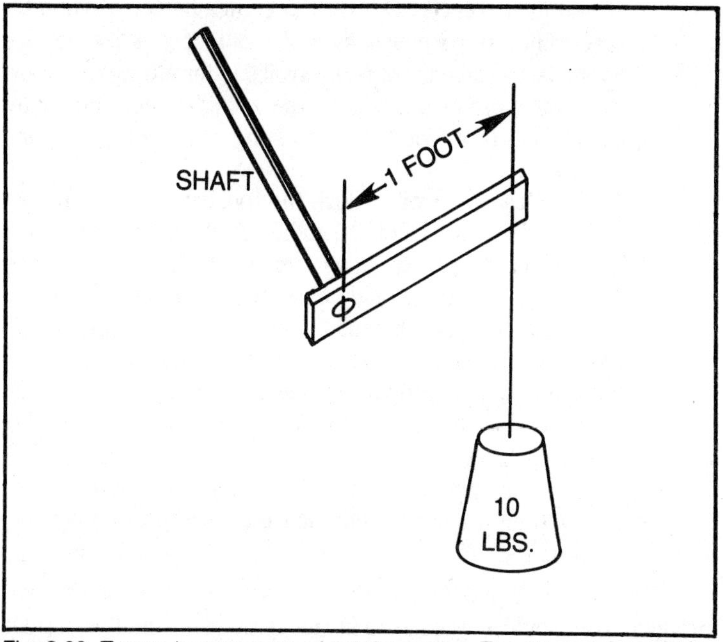

Fig. 2-36. Torque is a measure of twisting power. The 10 pound weight is applied at a distance of 1 foot from the shaft resulting in a torque of 10 foot-pounds on the shaft.

actually changes direction within the wires of an electrical circuit. The cycles are the frequency at which this occurs.

HOW TO SELECT POWER TOOLS

It would be nice if you could tell everything you need to know about a power tool simply by looking at it. Unfortunately, a quality product and an imitation can look quite similar. The following rules will help you select quality tools:

- —Determine the tools you will use most often. Do this systematically by jotting down some of the projects that you are interested in completing. Then evaluate the tools you will need. For example, consider whether most of the parts come preassembled or if they must be formed from raw materials and shaped.
- —Consider your own level of experience in the use of power tools. If you are not at all familiar with their usage, consider purchasing a relatively inexpensive model rather than one equipped with every conceivable feature. This way, if the tool is damaged by misuse, the loss will be minimized. And,

as you gain proficiency, you will be in a better position to determine the particular kind of tool that you like best.

—Evaluate the amount of room available for storage. If your work and storage area is not large enough to accommodate a particular tool, you will be wasting your money by purchasing it.

—Examine the ease with which the tool can be used. If you find that the tool can not be handled comfortably or that you have difficulty in controlling it, you will be less likely to use it. Therefore, you should consider alternatives.

—Evaluate the ease of maintenance. For example, some tools are permanently lubricated when they are manufactured and require little or no upkeep. Others require regular inspection and care. If you are the type of individual who wants to have a tool available for immediate use when needed and you don't want to think about it between uses—then required maintenance is an important consideration.

Unfortunately, it is often difficult to determine the quality of a product. You would have to examine its components and even then unless you were familiar with engineering concepts and part construction you probably still wouldn't be able to recognize fine quality. Under these circumstances, reliance upon the manufacturer's reputation, guarantees and warrantees offered by the manufacturer and the store becomes important.

SAFETY IS NO ACCIDENT

Safety, the freedom from danger and injury, is an important design consideration for all reputable tool manufacturers. They would rather build safety features into their tools than rely solely upon the judgment and actions of people using their products. However, even the best of safety features can prove inadequate unless the tool is properly used by the individual. The following safety rules are good practices to be followed by every carpenter:

—Wear protective clothing. Besides guarding your regular garments from damage, protective items such as shop aprons, safety goggles, work gloves and solid shoes can help prevent injury.

—Avoid wearing loose clothing. Items such as ties or apron strings can easily be caught by tools.

—Never use any electric tool where gasoline or other combustible liquids or gases might be present.

—Never use any electric tool when standing on wet or damp ground unless there is adequate insulating material between you and the ground. Do not rely upon devices such as ground fault interruptors to prevent accidents.
—Be certain that the tool is properly connected to the power source.
—Check to be sure that electric tools are grounded properly.
—Place any power cords out of the way so that they will not be cut or damaged by equipment or interfere with the operation of any tool.
—Be sure that any power tools used are run at the proper voltage.
—Work only on material that is securely fastened or clamped so that it can't shift or rotate.
—Do not change bits, drills, blades, or accessories on power tools with the cord plugged into the outlet. Don't rely on power switches to remain in the off position while such changes are made.
—Guards and safety features are included on power tools for an important reason—use them.
—Pay attention to what you are doing. Lack of concentration, even with hand tools, causes accidents.
—Keep all your tools in good condition. Power tools need protection from dampness or dirt.
—When using a tool for the first time, be sure to read any directions or instructions that might be enclosed. Being familiar with the tool will not only help prevent mishaps, but also allows you to do a better job.

3 What You Should Know About Building Materials

Carpentry is concerned with the building of useful items. To do this successfully you will need a knowledge of tools and their use and the ability to select the best material for the job.

CHOOSING LUMBER FOR THE JOB

To some individuals, everything that comes from a tree is classified as wood. However, there are actually two broad categories of wood, *hardwood* and *softwood*. Although these terms are widely used, they can be very misleading. The terms actually relate to the type of tree that is the source of the wood rather than the wood itself. *Deciduous* or *broadleafed*, trees yield hardwoods. Softwoods come from trees called *conifers* or *evergreens*. In fact, some hardwoods are actually softer than some softwoods. For example, Poplar and Philippine Mahogany which are hardwoods are actually softer than Douglas Fir which is a softwood. Even Balsa, long noted for its light weight and fragile composition is actually a hardwood.

Hardwoods and softwoods generally have different working properties. For one thing, hardwoods are usually more difficult to work with. The higher costs often associated with these types and the difficulties of workmanship are compensated by the durability of hardwoods and their beauty. In order to obtain a quality result from carpentry projects, it is important to know the basic properties of wood and how to select the correct wood for the job.

The Basic Types of Lumber

Wood comes in various forms, and there is a special terminology for each one. One of the most common forms is *boards*. These are lumber, generally rough, that is less than two inches thick. If the wood is from 2 to 5 inches thick and at least 2 inches wide, it is referred to as *dimension lumber*. Wood having these dimensions and positioned with the wide face horizontally are called *planks*. One common use of planks is for flooring. Studs are lumber that is 2 inches by 3 inches or 2 inches by 4 inches that is used for framing within the walls of a house. *Strips* are wood that have relatively small dimensions, less than 2 inches thick and smaller than 3 inches in width. *Timbers* have dimensions of at least 5 inches and are used in heavy construction.

Certain kinds of wood are used for specialized purposes. For example, *paneling* makes a beautiful wall covering. Decorative *molding* provides the right finishing touch for paneling, wall paper, or other coverings and *shelving* is used for bookcases and some furniture items. *Shakes* are used to provide beauty and protection to the outside of a house, while *plywood* is used in making the house itself and many of the items found within it. A complete list of speciality items would be quite extensive.

How to Select Lumber

Choosing the proper wood for the job is really not difficult once you understand how lumber is graded and sized. Basically, lumber is graded according to the number and types of defects it possesses. Remember, wood is a natural product derived from trees, and trees do not grow perfectly. Even if a perfect piece of wood could be obtained, the aging, processing, shipping and storage that the wood undergoes could contribute to the possibility of defects.

Knots are one of the most common blemishes found in wood. Knots indicate the spots where branches once met the trunk of the tree. They are recognized by their generally dark appearance and spots of knurled wood. Although they are technically a blemish, knots can add to the warm appearance of the wood and, in fact, might be desirable for the effect they provide. They are a defect only if you do not want this appearance. Knots are illustrated in Fig. 3-1.

Splits are separations that run along the grain of the wood. When a split does not extend too deeply into the wood, it is referred to as a *check*. Depending on their seriousness, checks can often be ignored or filled in when many carpentry projects are constructed.

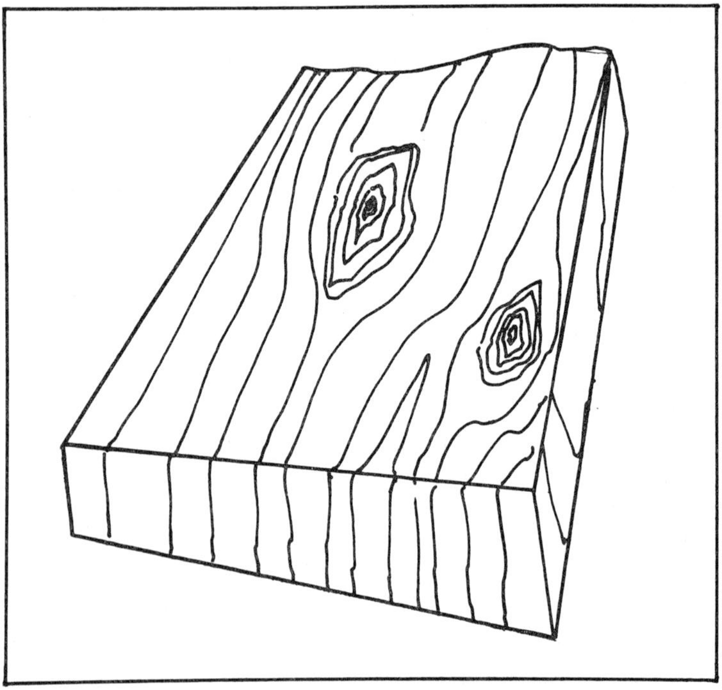

Fig. 3-1. Knots are formed where the branches of a tree met its trunk. They have a dark, knurled appearance and might actually come lose from the lumber.

Splits, however, generally require more extensive corrections and you might want to avoid this type of wood. The difference between checks and splits is depicted in Fig. 3-2.

Wane is either bark or a lack of wood along an edge or corner of the lumber. Depending on the nature of the project you are developing, you might choose to ignore this defect. Similarly, *pecks* (Figs. 3-3) probably will not cause you any difficulty. Pecks are caused by small pockets of decay within the tree and are really areas of disintegrated wood. However, as long as the lumber has been seasoned properly, pecks will not increase in size.

Shakes are defects that result from the separations between growth rings and they run along the grain of the wood.Note: Although the same term shake is used for roof shingles or siding on a house, the only relationship is that the siding consists of a rough split lumber. A wood shake in that sense does not refer to a defect.

Cupping is the curvature of the wood that you can see when the end of a board is examined. *Bowing* is a "dip" that can be observed when sighting along the surface of the wood. Whenever a wooden

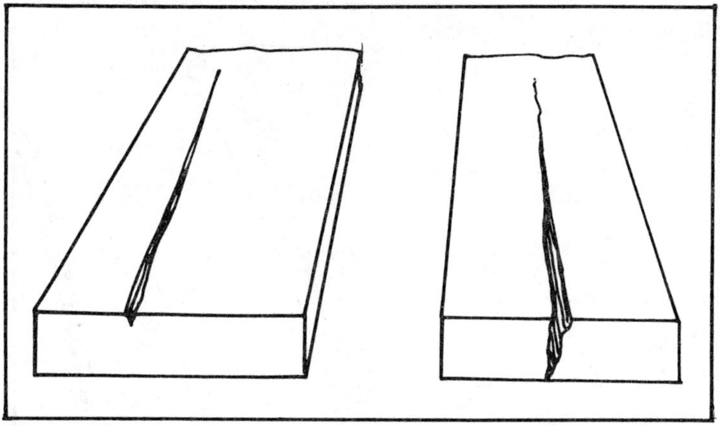

Fig. 3-2. Checks and splits occur when the wood separates along the grain. The separation in the board (left) is not as extensive as the one at the right—it is a check.

surface has a varies from being flat, it is *warped*. Cupping and bowing are two examples of warped surfaces (Fig. 3-4).

Lumber is usually sold either by the *board foot* or the *linear foot*, which is sometimes referred to a *running foot*. The number of board feet in a piece of lumber is obtained by measuring its length, width and thickness (all in feet) and then multiplying these numbers together. For example, a piece of wood that is 12 feet long, 2 feet

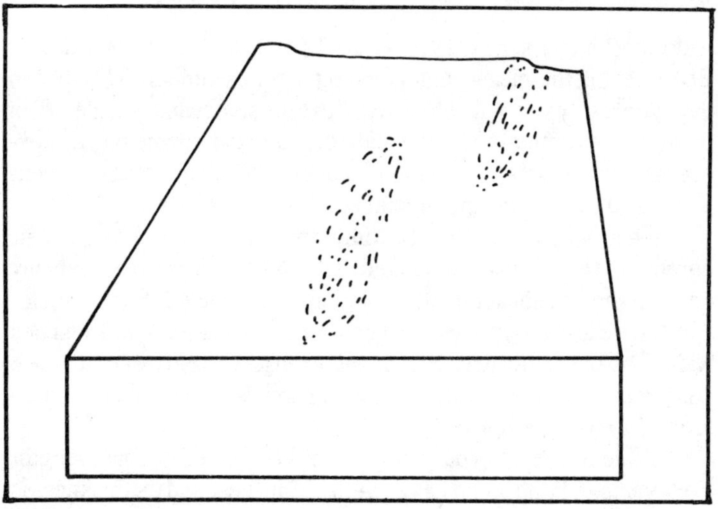

Fig. 3-3. Pecks are caused by areas of decay in a living tree. Once the wood has been seasoned properly they do not cause problems.

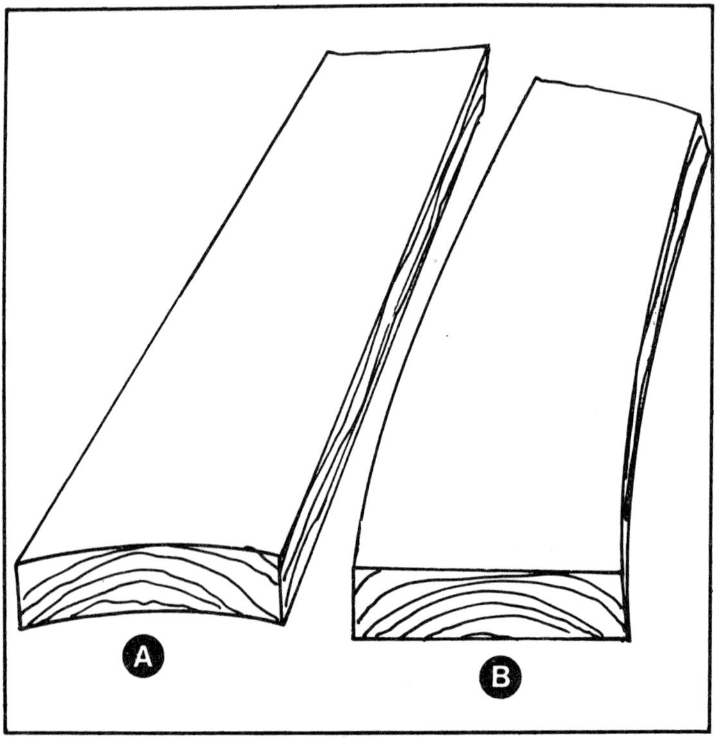

Fig. 3-4. (A) cupping can be determined by examining the edge for curvature. (B) bowing is visible by sighting along the length of the wood.

wide and 1 inch thich (1/12 foot) has 12 × 2 × 1/12 or 2 board feet. Board feet is the measurement used for rough lumber. This means that you can expect finished wood to be somewhat smaller than these measurements would indicate. To avoid confusion, the expression *net board feet* is frequently used when board measurement is used to determine the amount of *finished lumber*.

When wood of the same width and thickness is being considered, the term linear foot is used. The "foot" referred to is actually one foot length, since the other dimensions are fixed. For example, 2 by 4s are studs that are approximately 2 inches thick and 4 inches wide. The only difference that might be observed between two such studs would be their length, or in other words, the number of board feet that each possesses.

Table 3-1 allows you to make rapid estimates of board length. For example, if the wood has a width of 3 inches, and its thickness is 1 inch or less than 1 inch, the number of board feet can be calculated by taking the length of the wood and multiplying it by one-fourth. As

Nominal size (in.)	Actual length in feet								
	8	10	12	14	16	18	20	22	24
1 × 2		1⅔	2	2⅓	2⅔	3	3½	3⅔	4
1 × 3		2½	3	3½	4	4½	5	5½	6
1 × 4	2¾	3⅓	4	4⅔	5⅓	6	6⅔	7⅓	8
1 × 5		41/6	5	5 5/6	6⅔	7½	8⅓	91/6	10
1 × 6	4	5	6	7	8	9	10	11	12
1 × 7		5⅝	7	81/6	9⅓	10½	11⅔	125/6	14
1 × 8	5⅓	6⅔	8	9⅓	10⅔	12	13⅓	14⅔	16
1 × 10	6⅔	8⅓	10	11⅔	13 1/3	15	16⅔	18⅓	20
1 × 12	8	10	12	14	16	18	20	22	24
11/4 × 4		41/6	5	55/6	6⅔	7½	8⅓	91/6	10
11/4 × 6		6¼	7½	8¾	10	11¼	12½	13¾	15
.1/4 × 8		8⅓	10	11⅔	113⅓	15	16⅔	18⅓	20
11/4 × 10		10 5/12	12½	14 7/12	16⅔	18¾	205/6	22 11/12	25
11/4 × 12		12½	15	17½	20	22 ½	25	27½	30
11/2 × 4	4	5	6	7	8	9	10	11	12
11/2 6	6	7½	9	10½	12	13½	15	16½	18
11/2 × 8	8	10	12	14	16	18	20	22	24
11/2 × 10	10	12½	15	17½	20	22½	25	27½	30
11/2 × 12	12	15	18	21	24	27	30	33	36
2 × 4	5⅓	6⅔	8	9⅓	10⅓	12	13⅓	14⅔	16
2 × 6	8	10	12	14	16	18	20	22	24
2 × 8	10 2/3	13 1/3	16	18⅔	21⅓	24	26⅔	29⅓	32
2 × 10	13⅓	16⅔	20	23⅓	26 2/3	30	33⅓	36⅔	40
2 × 12	16	20	24	28	32	36	40	44	48
3 × 6	12	15	18	21	24	27	30	33	36
3 × 8	16	20	24	28	32	36	40	44	48
3 × 10	20	25	30	35	40	45	50	55	60
3 × 12	24	30	36	42	48	54	60	66	72
4 × 4	10⅔	13⅓	16	18⅔	21⅓	24	26 2/3	29⅓	32
4 × 6	16	20	24	28	32	36	40	44	48
4 × 8	21⅓	26⅔	32	37⅓	42 2/3	48	53⅓	58⅔	64
4 × 10	26⅔	33⅓	40	46⅔	53⅓	60	66⅔	73⅓	80
4 × 12	32	40	48	56	64	72	80	88	96

Table 3-1. Nominal size and length to board feet.

Table 3-2. A rapid calculation of board measure.

Width	Thckness	Board feet
3″	1″ or less	¼ of the length
4″	1″ or less	⅓ of the length
6″	1″ or less	½ of the length
9″	1″ or less	¾ of the length
12″	1″ or less	Same as the length
15″	1″ or less	1¼ of the length

a practical example, suppose that you wish to determine the number of board feet in a piece of lumber 3 inches wide, three quarters of an inch thick and 12 feet long. The number of board feet is obtained by taking 12, multiplying it by one-fourth to arrive at 3 board feet. The chart in Table 3-2 shows similar calculations for wood up to 15 inches wide.

Table 3-2 can be used to determine board feet for several different sizes of lumber. To use the table, first find the width and thickness of the wood in the nominal size column. Place a straight edge along the row corresponding to this entry. Next, locate the length of the lumber near the top of the chart. Move your finger down the column corresponding to this length until the straight edge is encountered. The number shown is the number of board feet. For example, a piece of wood that is 2 inches by 4 inches and 12 feet long would have 8 board feet.

HOW PLYWOOD IS CONSTRUCTED

Plywood is one of the most popular and useful wood products. Construction, panelling, furniture, cabinets and shelves are just few of the many uses for this product. Despite this, many people don't know the basic facts about plywood and are unable to make full use of its features.

In addition to being versatile, plywood is one of the strongest wood items. Its strength is derived from the way it is constructed. If you examine plywood you will see that it really isn't made of a solid sheet. Instead it consists of several layers. Each layer is glued to the next one so that the grain of one layer is at right angles to that of the next layer. An odd number of layers is used.

What distinguishes one type of plywood from another? The answer is the type of glue used to bond the layers and the grade of these layers. Plywood used for exterior purposes has a type of glue that is extremely resistant to moisture and heat. While some interior plywood can be purchased that has this type of bonding, the grade of the wood will be different.

Table 3-3. Recommendations for nailing plywood siding directly to studs.

Plywood Siding			Max. Stud Spacing (inches)		Nail Size (Use nonstaining box, siding or casing nails)	Nail Spacing (inches)	
Type	Description	Nominal Thickness (inches)	Face Grain Vertical	Face Grain Horizontal		Panel Edges	Intermediate
Panel Siding	MDO EXT-APA	19/32, 3/8	16	24	6d for panels ½" thick or less; 8d for thicker panels.	6	12
		1/2 & thicker	24	24			
	303–16 o.c. Siding EXT-APA (including T1-11)	5/16 & thicker	16	24			
	303–24 o.c. Siding EXT-APA	7/16 & thicker	24	24			
Lap Siding	MDO EXT-APA	19/32, 3/8	–	16	6d for siding 3/8" or less; 8d for thicker siding.	4" @ vertical butt joints; one nail per stud along bottom edge.	8" @ each stud, if siding wider than 12".
		1/2 & thicker	–	24			
	303–16 o.c. Siding EXT-APA	5/16, 19/32, or 3/8	–	16			
	303–16 o.c. Siding EXT-APA 303–24 o.c. Siding EXT-APA	7/16 & thicker	–	24			

Plywood is available in various dimensions; the most common sheets are 4 feet by 8 feet and one-half inch thick. However, if the job calls for it, you can get a thickness of from one quarter-inch to one and one eighth-inch. Widths usually run from 3 to 5 feet and lengths can be obtained up to 12 feet.

The type of plywood to get depends on the job to be performed. If you are planning to use it for panelling or furniture, hardwood types are best. The softwood variety is the type used for most carpentry projects. See Table 3-3 for how to use plywood as siding.

Plywood is available in various grades. Hardwood # 1 is an extremely high quality product and it is expensive. The #2 grade is not matched quite as carefully as#1 type and small knots might be present. The #3 grade is adequate when the final product will be painted.

Softwood is coded A, B, C, D and N. For a quality finish, the N (natural) grade is the one to select. Otherwise, the A grade is better quality than B, and so on. Keep in mind that these letters are really general guidelines. The lower grades are not necessarily an inferior product. However, if the plywood is to be used in a project where it won't show or where it will be painted or a special effect is to be achieved, the lower grade product might be adequate.

Plywood is available as a finished or an an unfinished product. While the finished type costs more, the additional expenditure might be worthwhile since you will avoid the additional labor and time to finish them yourself. In addition there are specially finished products for decorative use. These are available in a wide assortment of colors and patterns ranging from natural woodgrain to marble. Their cost, of course, varies with the quality of the finish.

Because of the special construction of plywood, it must be handled in a slightly different manner than other wood. For example, because the outer layers of the product are thin, they have a tendency to splinter as the wood is sawed. It is important to use a saw blade having a large number of small, sharp teeth and to conrol the rate of cut. Cutting too fast will promote splintering. If possible, use a special plywood cutting blade with a power saw or a hand saw with a 10 point blade. If it is necessary to use a regular power saw blade, make sure that the blade extends through the wood. This cushions the impact of the teeth on the plywood.

The side the saw blade enters the plywood also is important. If the blade is allowed to enter the wood on the "good" side (the side that will be visible) any splintering that occurs as a result of the sawing will occur on the back as the blade exists the plywood. If you decide to cut the plywood with a portable circular saw, place the plywood so that the good side is down. Splintering can be reduced by placing a piece of scrap wood along the line of the cut where the blade would normally exit from the wood.

One other difficulty encountered when working with plywood is sagging. Plywood sheets are flexible and unless they are properly supported they will bend. This bending sometimes makes it difficult to properly cut plywood sheets.

Although careful cutting of plywood results in a somewhat smooth edge, additional effort may be necessary. If the edge is not even, or the dimensions of the wood are slightly larger than needed, plywood can be carefully planed and sanded.

How to Plane Plywood

- Set the plane for thin cutting.
- Work from the ends toward the middle of the plywood sheet. This helps to avoid chipping at the ends.
- Use long, smooth cuts.

Don't worry in the event some minor splintering occurs when you saw or plane plywood. Small irregularities or rough spots can be removed by sanding.

How to Sand Plywood

- Do not sand the surface of a plywood sheet unless the surface is being repaired or it is absolutely necessary. Remember, the layers are thin and overworking the surface will wear it away.
- Sand along the direction of the edges.

HOW TO TREAT PLYWOOD IRREGULARITIES

Because of the special construction of plywood, treating irregularities or slight damages requires some special considerations.

Filling Deep Scratches and Smallholes

— Use a quality wood putty.
— Use a putty knife to apply the putty slightly higher than the surface of the panel being prepared.
— Let the putty dry thoroughly.
— Using a fine sandpaper and light pressure, carefully sand the repaired spot until it is smooth and even with the wood surface.

Finishing a Plywood Surface

In order to obtain a special effect or to cover many irregularities in the surface of the plywood, you might decide to apply a finish. If additional sanding is required:

— Sand along the grain.
— Use a sanding block to provide even pressure and help prevent wearing through the finish.
— Use a fine sandpaper.
— Apply only light pressure.
— Sand in the direction of the grain.

FASTENING PLYWOOD

Plywood is usually fastened to other materials with nails. When additional holding strength is desired, holes are drilled into the panels and then flat-head screws are used. Additional holding power can be obtained with the use of glue. The type of glue used depends on whether the plywood is being used indoors or outdoors.

Unlike some materials, plywood can be nailed near its edges without splitting the wood. However, nailing into the edges themselves is tricky.

Some of the inexpensive grades of plywood paneling are extremely thin. These poorer quality sheets are often difficult to manage. Their high degree of flexibility makes the job of accurate cutting tricky and splinters or actual splitting often result from finishing or smoothing. Occasionally, even screws will not hold this grade of plywood; gluding is about the only alternative.

STORING PLYWOOD

Ideally, plywood should be used shortly after its purchase. However, there could be times when you will not be able to complete a project quickly. If the project is somewhat extensive, you might not use all the plywood at once. Under these circumstances, it is important to store the plywood sheets carefully to avoid them.

- Store the panels in a cool, dry area. This is important since moisture will damage the finish of some types of plywood, particularly if the edges are permitted to be exposed.
- Position the panels so that they do not come in contact with the ground, but are permitted to lie flat.
- Support the plywood sheets so that they do not sag.

- If it is necessary for the wood to be covered, be sure that there is sufficient air circulation beneath the cover. Otherwise, moisture condensation can be a problem.
- When positioning the plywood for storage and when moving it to the location where it will be used, be particularly careful to avoid scratching or damaging the surface of the panel. Remember that damage to the layered edges might be a problem, so additional care with them is also worthwhile.

HOW TO STORE LUMBER

Storing wood improperly ruins it, costs you money and can even be dangerous. A little care is all that is needed to protect your investment.

— Try to avoid storing lumber whenever possible. Plan your projects carefully so that large amounts of wood are not left over. Purchase what you need but allowing for a little waste. Use the lumber as soon as possible.
— Store wood in a cool, dry place.
— Don't allow boards to be supported only by the ends since they will tend to sag in the middle. Place wood flat whenever possible.
— Keep wood away from heat sources or open flame.
— Store chemicals away from the area where you keep the wood.
— Allow air to circulate around the wood. Don't let covers lock in moisture.
— Don't store other items on top of the wood.

MOLDINGS

Moldings are those strips of wood used to finish windows and doors, and to separate walls from floors. They are sometimes placed between ceilings and walls and in some cases, they are even are located in the middle of a wall. They can be used to provide an extra decorative touch and supply that "finished" look.

But carpentry craftsmen like molding for a different reason. It can be used to hide inaccuracies and errors while at the same time appearing to be part of the original project design. By selecting from the wide assortment of moldings available, you might even be able to give the appearance of entirely redecorating a room.

Molding Tips

Basically, there are two types of moldings—those applied to flat areas such as doors or windows, and curved moldings that are usually decorative or used for finishing. An example would be the molding placed where the floor meets a wall. Each type, although similar in purpose, must be handled in a slightly different manner.

Flat molding is usually joined together at 45° angles. If you have a miter box, these angles are easy to cut. However, the molding can still be cut properly without a miter box if care is taken to measure and cut the molding at the proper angle. Accurate measuring can be made by using one of the angle layout tools discussed in chapter 2.

Curved moldings and some fancy, ornamental types require some additional attention when they are joined. Not only must the angle at which they meet be considered, but they must have matching curves as well. The best way to match the curves is by using a coping saw.

The easiest way to attach molding is to use finishing nails.

How to Attach Molding

The nailset technique:

- Place the molding in position.
- Using a tapping stroke, position finishing nails with a hammer.
- Carefully drive the nails. Avoid marring the molding.
- Drive the heads with a nailset.

The blind nail technique:

- Use a small knife to create a small sliver of wood. This sliver will be used to hide the nail. Be careful to avoid breaking the sliver from the wood.
- Repeat the first step along the length of the molding before any nails are actually driven.
- Gently move each sliver so that its gouge is exposed and drive a finishing nail.
- Glue each silver back into place.
- Rub each area lightly with a piece of fine sandpaper.

The advantage of the blind nail procedure is that all traces of fastening are hidden. However, it does take more time and care to accomplish than the nailset technique.

HOW TO ENSURE EXCELLENT WOODCUTTING

There is no such thing as "perfect" woodcutting, because wood itself is not perfect. Imperfections provide much of the beauty and warmth of wood. What is possible, however, is the appearance of a perfect job. Attaining the look of perfection is an ability anyone can master.

- Plan! Decide what you want to do, where you are going to do it and how it will be done.
- Use the proper materials for the job. Don't expect inferior materials to result in a quality product. This does not mean that only expensive materials can be used. It does mean that materials of sufficient quality to achieve the desired results should be used.
- Use tools carefully. Extra care can overcome inexperience when many tools are used.
- Don't hurry. Although you might be anxious to see a job finished, speed leads to errors and is a safety hazard.
- Mark all pieces carefully and follow these measurements as closely as possible. This will assure that all items fit properly.
- When using paint, varnish, glue or other products, be sure to follow the directions that come with the package. Don't let impatience force you to allow insufficient time between applications, or tempt you into using the item before it is ready.
- If a minor mistake is made resulting in a gouge or a joint that is a little uneven, the use of a quality wood filler can often correct the mistake. Apply the filler according to the directions on the can. Allow it to dry and then lightly sand it until it is even with the wood. If you paint, the error won't show.
- Clamp delicate wood before cutting it. This helps prevent the wood from splitting when the saw blade is about to complete the cut.
- Don't put stress on joints that have been cut until they have been completely finished. This will prevent them from accidentally separating.

THE PURPOSE OF PLANING

Sometimes, despite a careful attempt to cut wood perfectly, the cut will not be absolutely true. There could be some uneveness along an end or edge or a cutting line might not have been followed exactly. When this happens, the excess or unwanted wood must be removed. If the amount of excess wood is extremely small, it can be sanded until the excess is removed. However, if the wood has noticeable uneveness or there is more than a very thin layer to be removed, sanding becomes too time consuming and might not give the desired results. Under these circumstances, planing is recommended.

Planing is often useful after a job has been completed. For example, a wooden door might have been installed perfectly in a new house or room. As time passes, a house can settle. That is, it shifts slightly as the ground adjusts to its weight. Stresses and strains can cause the door frame to become slightly out of line and no longer closes properly. If only a minor adjustment is necessary, the door frame or molding can be adjusted so that the door works properly again. However, it is sometimes easier to adjust the door itself, particularly when the wood has warped. While sanding is sometimes effective, planing the door is usually the best answer.

Although you specify the exact measurements of a piece of lumber purchased at a lumber yard, the wood you receive may not be accurately sized. Usually, there is not enough excess wood to cut away with a saw. Under these circumstances, using a plane is the best way to remove the extra wood.

Once you gain experience using a plane, you can use this tool to achieve special effects. For example, by holding the plane on a slight slant and using it along the outside edge or end of the wood, a *beveled*, or slanted, effect will result. When used properly, beveling

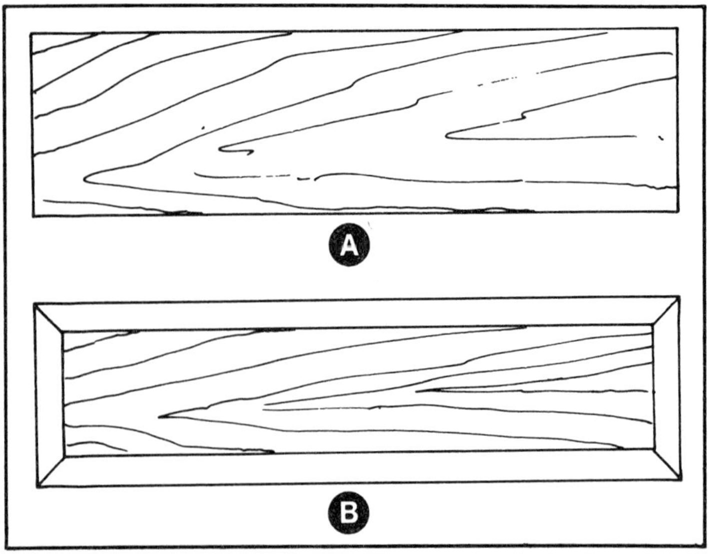

Fig. 3-5. Beveling is a simple process that can add a nice touch to a job. A represents a top view of a piece of wood. B shows that same wood after it has been beveled.

can make a simple item appear fancier and provides an extra touch to the final product. Some experience in the use of the plane is recommended, however, before this technique is applied because care must be taken to prevent splintering or splitting of the wood at the ends. Once you know how to use a plane, practice beveling on some scrap. The basic technique is not difficult and with a little practice professional-looking results can be achieved. Figure 3-5 illustrates the beveling effect.

THE IMPORTANCE OF SANDING

Sanding is probably the single most important step in wood finishing. Its purpose is to remove scratches, blemishes, and construction markings before a finish is applied and to lightly roughen a surface between coats of paint or other finishes so that they will adhere better.

The need for sanding can't be emphasized too strongly. Some people skip this step since they feel that staining or painting will cover the blemish in the wood. Actually, the opposite will sometimes happen. The rough spot is emphasized by certain finishes. Whenever a quality product is to be produced, carpentry craftsmen will actually sand three times before a finish is applied. The first, or rough sanding, is intended to remove surface defects or dirt and is generally done across the grain. The intent is to remove material quickly and uniformly. Care must be taken to avoid removing too much of the surface, particularly if a fine, thin wood is used.

Once the surface has been cleaned by the rough sanding, a second step of sanding is performed. This time a medium grade sandpaper is used and sanding is done along the grain through the use of either an electric finishing sander (Fig. 3-6.) or a sanding block. This block is nothing more than a flat piece of wood or metal that is wrapped with the sandpaper so that pressure can be applied to the surface uniformly. The purpose of this medium sanding is to remove the scratches and rough cuts caused by sanding across the grain with the rough sandpaper. After this step the product is often smooth enough to paint with good results. However, the individual who wants to add additional quality to the job will perform one more step—hand sanding.

Hand sanding a piece of wood is analogous to buffing a car after it has been waxed. The job will look good without it, but the individual who wants the most from his labors will always do it. Hand sanding is performed by pressing the fingers of one hand against a fine sandpaper and sanding in the direction of the grain. Since the wood has already been smoothed by the previous sanding steps, it isn't necessary to apply a great deal of pressure. A light touch produces excellent results.

Sanding can also be done when the wood is only slightly too large. Otherwise a plane would be used to shave the wood to the proper size. It is also a good idea to lightly sand the wood along the edges where they meet the surface. This will remove any small splinters that might have resulted from sawing or working with the wood.

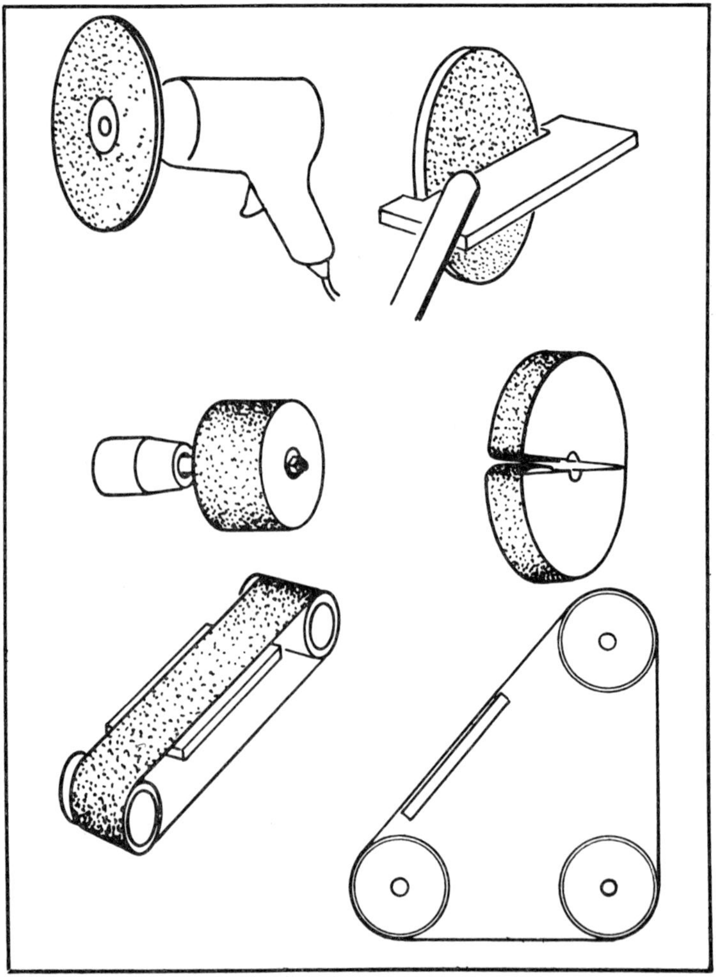

Fig. 3-6. Sanding disks are useful for shaping and finishing.

There is one other time when sanding is recommended, The manufacturers of many finishes state in their directions for using their product that sanding should be done between each application when many coats are being applied. This sanding is actually hand sanding and it allows the coat to adhere to the previous coat more easily. This produces a professional-looking result.

HOW TO CORRECT MISTAKES

Despite careful attention to detail and skill in the use of tools, even experienced carpenters make mistakes occasionally. However, the only way to correct some of them is to start again. Many errors can be corrected and often will not be noticed by those who did not know the mistake was made. Here are some things to consider:

— If the problem is a dent or deep scratch in the wood, do not try to remove it by using a plane or by doing a large amount of heavy sanding unless you have wood to spare. Instead, fill the dent with a quality wood filler. It is available in different shades to closely match the color of the wood. Natural color is generally the most convenient. Once the filler dries, it can be sanded smooth. After it is finished with stain or paint, the spot will blend with the surrounding area.

— Small spaces in joints can be filled by using the previous step. However, careful sanding is required so that the shape of the joint is as intended. If the space if too wide, the joint might not be strong enough for the project. Replacing a section may be necessary.

— If a length of wood turns out to be too short, extra wood can be added. If you are working with molding, simply make a miter cut where the additional wood is to be added to the short piece and nail both pieces to the wall. When they are painted, the two pieces placed together will look like one solid section. Other types of wood can be joined by constructing special joints. Unless these are preconstructed so that only positioning is required, they can be tricky to make and are best left to those with experience.

— Unwanted holes can also be filled with wood filler. However, "backing" is usually necessary. To do this, clamp a piece of scrap wood against the hole on the side of the wood that won't show. Then fill the hole as the first step. If the hole is deep, fill it in several stages. Apply a little filler, let it dry and then apply additional material. Repeat this process until the hole is filled completely, sand and finish.

— To hide minor blemishes in unusual wood, take a scrap piece and make sawdust. Mix the sawdust with filler and apply. After light sanding, the blemished area should blend with the surrounding wood.

HOW TO CONVERT MISTAKES INTO ASSETS

With a little imagination the mistake that looks so terrible to you when it is made can actually be used to your advantage. For example, suppose that the wood you prepared has an edge that was damaged or a joint that really doesn't mesh properly. See whether a piece of molding could be used to cover the mistake. Fancy molding generally looks as if its use was planned and adds a finished effect. Or perhaps a piece of metal, such as brass or aluminum, could be cut into a pattern and nailed over each joint to provide a different look. When finished properly, metal additions can provide unusual and attractive results.

If several flaws in the surface of the wood develop while a project is being constructed, try emphasizing them. Sometimes an interesting pattern develops that is not only unusual, but attractive as well.

If the problem is a blemish that shows through the final finish that is applied, consider using a fancy decal to cover it. These are thin patterns that stick to the wood and they can be quite attractive. For example, if the project is a baby's dresser, a decal of a favorite cartoon character would not only fit in with the decor of a child's room, but it would most likely appeal to the child more than the most perfectly finished furniture available in a store.

When constructing furniture, you might find that there are too many blemishes to cover. If so, consider *distressing* the wood. This is a process that is often used by furniture manufacturers. The wood is intentionally marred by tools and finished unevenly so that the final result is a product that has an aged or antique look. When this effect fits into the decor the results can be quite attractive. It is easy to get this effect. Simply hit the wood a few times with a hammer, lightly scratch it with a screwdriver or awl or tap it a few times with the corner of a metal ruler or similar object. Use your imaginaton for other ways to distress the wood.

If the finish on a piece of furniture doesn't turn out the way you like and you don't want to strip it and refinish the item, try antiquing it. There are many fine antiquing kits on the market and they are available at most paint stores. Most involve painting the wood, right over the finish that you don't like, with a dark base coat. Then, a special coat is applied to simulate a wood grain.

There are many other ways to turn mistakes into advantages. All that is required is the imagination and creativity that every carpentry craftsman possesses. Don't be afraid to experiment. Its fun!

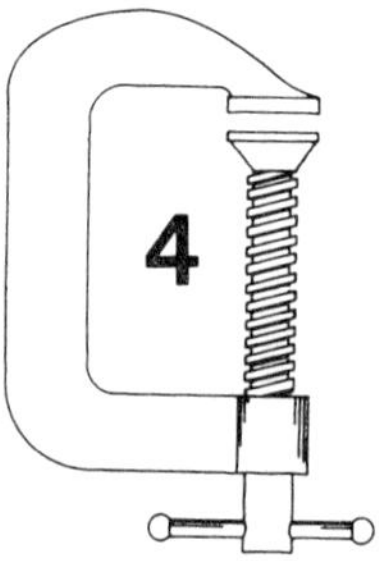

4 There Is Nothing Mysterious About Selecting Hardware

With the wide assortment of hardware that is available, it is not surprising that many people are confused about the proper items to use. Once you are familiar with the types of hardware available and their purpose, making the proper selection is easy. The trick is to understand the way each piece of hardware works and to determine which item best suits your needs.

THE DIFFERENCE IN NAILS

Originally, metal nails were small iron pins or spikes. Then, someone got the idea to put a head on them so they would be easier to position. Today, the simple nail has evolved into not one item, but a wide variety of fasteners made of different types of metals with great holding power. They are easier to use then their predecessors and look better as well. There was a time when nails were either homemade or built by a specialist. Today hundreds of different varieties are available at hardware stores, department stores, builder's supply houses or neighborhood lumber yards.

Common and *box nails* are the types most frequently used. They are very similar in construction. However the box nail is thinner, so some care must be taken not to bend it when hammering it into position. The tendency to bend if not struck properly is offset by the advantage that it is less likely to split the wood being fastened.

Casing and *finishing nails*, as their names imply, are used for finishing work. The finishing nail leaves a small hole that can be filled so that the nail is hidden when the project is completed. There are finishing nails that come in colors, that eliminate the need for this small amount of filling. Simply choose the color that matches the wood or paneling you are using. Casing nails are similar to the finishing type; however, they are heavier and stonger.

Sinkers are rosin-coated nails that are used when installing gypsumboard walls and ceilings. They offer additional holding power over regular bright nails. Since no one wants the head of a nail to be visible on a new or repaired wall, these nails come with a slight bevel on the bottom of the head so that they can be driven flush with the gypsumboard surface. This saves time that would have to be spent with a nailset and filler. This type of nail can be used for additional purposes when its special features are desired. For example, it is excellent for attaching rubber door bottoms to garage doors and with some types of rough molding.

There are *tacks* for laying carpets and attaching weatherstripping, a well as general purpose usage. These are small nails with wedge-shaped harp points that make them ideally suited for fastening.

There is even a nail that is meant to be removed! Called a *scaffold nail*, it has two heads and is intended for temporary structures. The nail is driven so that it's lower head is against the surface of the building material. The second head allows the nail to be gripped by a hammer, pliers, or pulling tool for easy removal of the nail.

If the carpentry project involves items to be left out doors for any length of time, for example, lawn furniture, special nails should be used. Ordinary nails will rust over a period of time when exposed to moisture. This is true even when galvanized steel nails are used since the coating on the head is often damaged by the hammer itself. Nails made of stainless steel or aluminum are probably your best bet here. However, care must be taken when driving aluminum nails since, even though new alloys have reduced the problem, they have a tendency to bend. Aluminum nails are generally less expensive than stainless steel nails. However, the stainless variety works better when hard wood is involved.

There are specific nails that are recommended for use in roofing. The proper nail to use depends on the roofing material being used. If you are repairing or installing a roof, inquire about the best nails to use when you purchase the roofing material.

Shingle nails, as the name implies, are intended for use in installing roof shingles and shakes and when new shingles are to be placed over old ones. They come in a variety of lengths and it is best to ask about the best nails to use when you purchase the shingles. Sometimes, small box nails can be used if only one or two shingles have come loose or need replacing.

So far, I have only discussed nails used to fasten wood or building materials. However, there might be times when you wish to fasten an item to a masonry such as concrete, stone, concrete blocks or brick. There are special *masonry nails* for this available in different styles and sizes. A word of caution. Masonry nails are fine for holding items so that the item will not move up or down or sideways. However, they sometimes have little holding power if an outward force is applied to the item being fastened. In such cases, there are special masonry fasteners that should be used instead of masonry nails.

For projects such as installing floors, *cut nails* are frequently used. They are wedge shaped and have a head on the large end. This type of nail is made of hard steel and is noted for its good holding ability.

There are specialized nails for fastening insulation board, canvas, plastic, plywood and other building materials. There are *pin nails* used to fasten thin materials such as ceiling tile or thin wood, *brad nails* for delicate or finishing work and large nails for construction jobs. They are available in smooth shanks or threaded varieties. The screw-shanked nails provide greater holding power than the smooth type. The list of specialized nails is too extensive to give a complete accounting here. If special fastening is required for a project, ask your hardware supplier about the correct nail for the job. Figure 4-1 shows some of the more commonly used types of nails.

Just as there is a wide variety of nails, there is also a large selection of sizes. Nails are measured using the term *penney*. At one time, one hundred hand-made nails could be purchased for a few pennies. The term has evolved from the actual cost to the size of the nails. Most hardware stores carry a wide assortment of nails in 2-penny to 16-penny sizes. However, some nails are available in larger sizes. The abbreviation for penney is d. You will often see this when nails are sold by weight from open bins.

When a nail is larger than 60 pennies it is referred to as a *spike*. Rather than being measured by pennies, these fasteners are distinguished by their diameter and length. Usually, spikes are used in

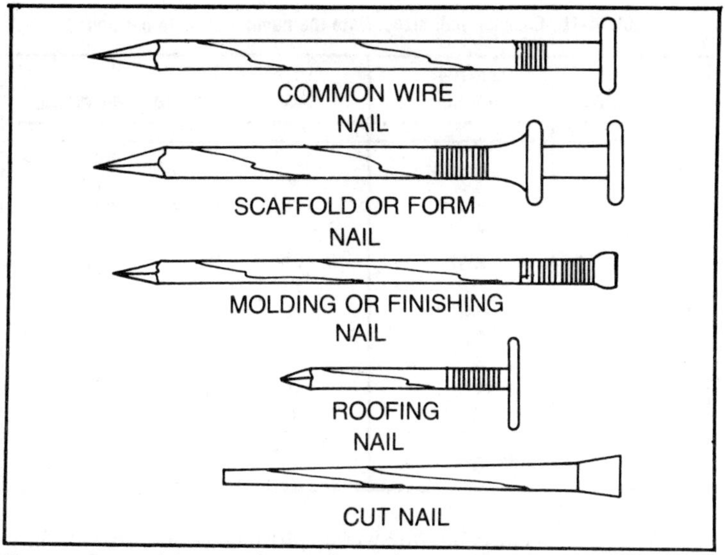

Fig. 4-1. Some of the most frequently used types of nails.

heavy construction projects and you probably won't use them often in carpentry craft projects.

Tables 4-1 and 4-1A shows the sizes of nails. Each size is related to its diameter in inches and the corresponding steel wire size (gauge) that is used to form the nail. The bottom line shows the quantity of each size nail that would weigh one pound.

NAILING TIPS

It is generally better to drive the nail into the wood at a slight angle, rather than straight into it. This is called *toe-in* and increases the holding power of the nail. This is shown in Figure 4-2. When several nails are used, reverse the direction of the slant if possible as each additional nail is driven. An example of this is shown in Figure 4-3.

Table 4-1A. Nails Come in a Variety of Sizes.

Lengths (inches)	Diameters (inches)			
	3/16, 1/4 5/16, 3/8	7/16, 1/2	9/16, 5/8, 3/4	
¾	x			
1	x	x		
1¼	x	x	x	
1½, 2, 2½, etc., 9½, 10 to 20.	x	x	x	x

Table 4-1B. Common nail sizes. Note the number of nails per pound.

PENNY SIZE	LENGTH (INCHES)	GAGE	NUMBER PER POUND
2	1	15	840
3	1¼	14	540
4	1½	12½	300
6	2	11½	160
8	2½	10¼	100
10	3	9	65
12	3¼	9	65
16	3½	8	45
20	4	6	30
30	4½	5	20
40	5	4	17
50	5¼	3	14
60	6	2	11

Common sense is the rule when determining the proper length nail to use for a particular job. However, a general rule to follow is that the length of the nail should be approximately three times the width of the item being fastened. This will provide the strongest holding ability, assuming the proper type of nail is selected. Approximately one third of the length holds the wood being fastened, while two thirds of the length is driven into the second piece for anchorage.

If holding ability is a major consideration for a particular project, consider threaded nails. They have considerably more holding power than conventional types and they tend to maintain this ability for longer periods—particularly as the wood starts to age.

When using nails, try to position them across the grain. Nails driven in the direction of the grain do not hold as well as those positioned across the grain.

Using a large number of nails does not necessarily increase the holding ability. It is better to use only a few nails of the correct type and size because properly driven nails will hold better than a great many nails driven close together.

Table 4-2 shows different sizes and types of nails and lists how they are used.

WHEN TO USE STAPLES

Staples are u-shaped wires that are used for fastening a large variety of objects that range from wire, to ceiling tiles, to up-

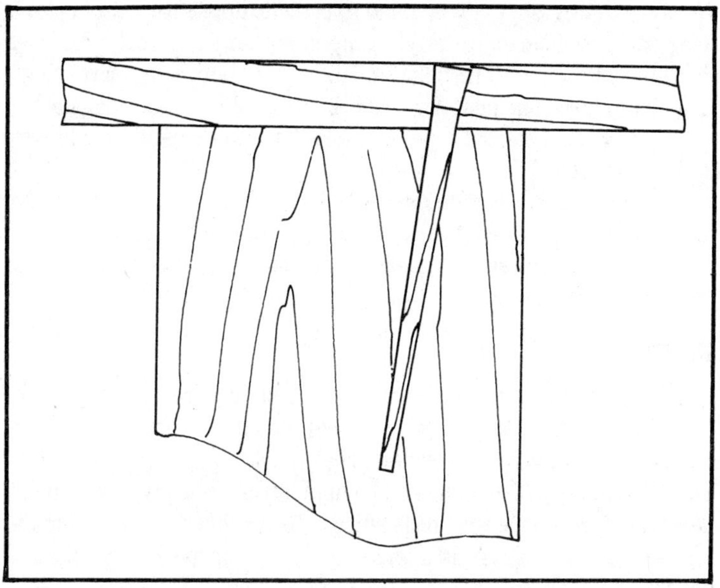

Fig. 4-2. Positioning the nail so that it has a slight angle compared with the surface of the wood increases the holding power of the nail.

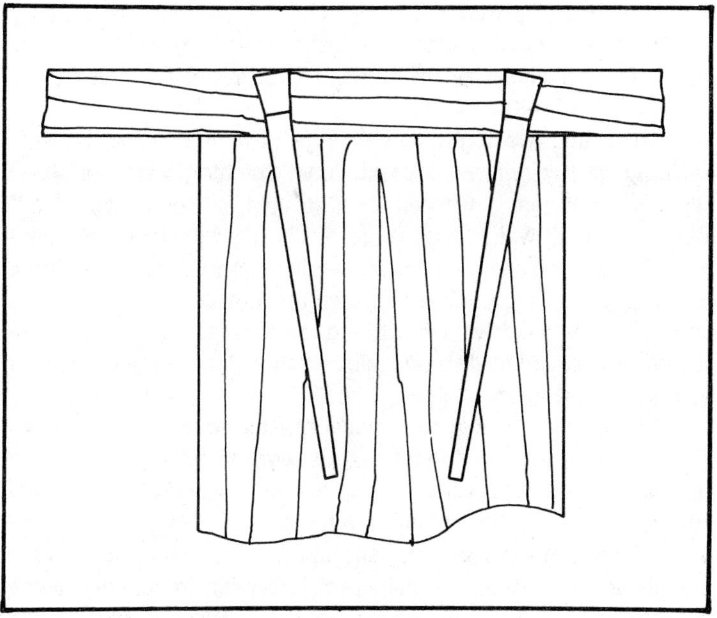

Fig. 4-3. Reversing the direction of the slant when several nails are to be driven also increases the holding ability of the nails.

holstery. They can even be used to provide hooks for keeping your carpentry tools in an orderly arrangement on your work area wall. Because of their easy installation, staples are becoming increasingly popular. This is one fastener that is limited only by your imagination. There is even a special type of staple that can be used to hold wire fence to a post.

Staples can often be used instead of nails for projects where particularly light holding ability is required or when the staples won't show. Be sure to ask the building material supplier whether staples are recommended for a particular project.

SCREWS

Screws are more expensive than nails and generally require more time to position. They are sometimes necessary for superior results and should be used when greater holding power is required. Screws can easily be tightened to hold items securely, are usually neater in appearance than nails and can be removed without damaging the material. There is a wide variety of screws designed for a particular purpose.

Common *wood screws* are generally made of unhardened steel, stainless steel, brass or aluminum. Those that are made of steel might be blued, plated with zinc, chrome cadmium or brightly finished. They have a thread running about two-thirds of their length from a gimlet point. A head with a slot for use with a screwdriver caps the screws.

The head style determines the type of wood screw. *Flathead, ovalhead* and *roundhead* are the most common forms of wood screws. Each comes with either a straight slot or Phillips head. When you don't want a flathead or ovalhead screw to show, countersink the wood and then use a covering material such as a plastic wood or dough. Roundhead screws are not countersunk. They should be driven firmly flush with the wood surface. It is a good idea to position a roundhead straight slot screw so that the slot is parallel with the wood grain.

Lag screws are longer and heavier than common wood screws. They have a coarse thread for slightly more than half the length of the screw and a gimlet point. They are used when ordinary wood screws are not heavy enough for the job or when those screws would be too short. They are generally used in heavy duty work in conjunction with expansion anchors as in framing timbers to existing masonry. Hexagon and squarehead lag screws are usually positioned with a wrench.

Table 4-2. How nails of various types and sizes would be used.

Size	Lgth (in.)	Diam (in.)	Remarks	Where used
2d	1	.072	Small head	Finish work, shop work.
2d	1	.072	Large flathead	Small timber, wood shingles, lathes.
3d	1¼	.08	Small head	Finish work, shop work.
3d	1¼	.08	Large flathead	Small timber, wood shingles, lathes
4d	1½	.098	Small head	Finish work, shop work
4d	1½	.098	Large flathead	Small timber, lathes, shop work.
5d	1¾	.098	Small head	Finish work, shop work.
5d	1¾	.098	Large flathead	Small timber, lathes, shop work.
6d	2	.113	Small head	Finish work, casing, stops, etc. shop work.
6d	2	.113	Large flathead	Small timber, siding, sheathing, etc., shop work.
7d	2¼	.113	Small head	Casing, base, ceiling, stops, etc.
7d	2¼	.113	Large flathead	Sheathing, siding, subflooring, light framing.
8d	2½	.131	Small head	Casing, base, ceiling, wainscot, etc., shop work.
8d	2½	.131	Large flathead	Sheathing, siding, subflooring, light framing, shop work.
8d	1¼	.131	Extra-large flathead	Roll roofing, composition shingles.
9d	2¾	.131	Small head	Casing, base, ceiling, etc.
9d	2¾	.131	Large flathead	Sheathing, siding, subflooring, framing, shop work.
10d	3	.148	Small head	Casing, base, ceiling, etc., shop work.
10d	3	.148	Large flathead	Sheathing, subflooring, framing.
12d	3¼	.148	Large flathead	Framing, bridges, etc.
16d	3½	.162	Large flathead	Framing, bridges, etc.
20d	4	.192	Large flathead	Framing, bridges, etc.
30d	4½	.207	Large flathead	Heavy framing, bridges, etc.
40d	5	.225	Large flathead	Heavy framing, bridges, etc.
50d	5½	.244	Large flathead	Extra-heavy framing, bridges, etc.
60d	6	.262	Large flathead	Extra-heavy framing, bridges, etc.

[1]This chart applies to wire nails, although it may be used to determine the length of cut nails.

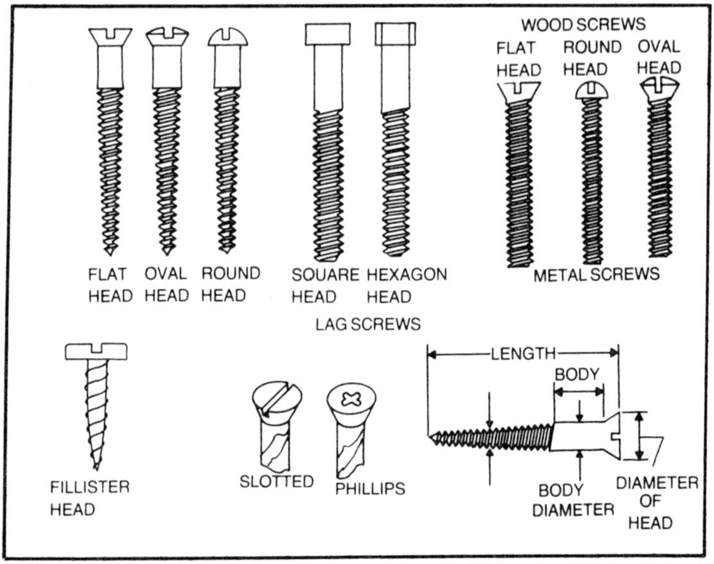

Fig. 4-4. There are many different types of screws. While most are positioned with a screwdriver, lag screws are similar to bolts and are tightened with a wrench.

Sheet metal screws are used to assemble metal parts. They are usually made of steel or brass and are available with four different types of heads: flat, oval, round and fillister. They are available in a wide assortment of sizes. Figure 4-4 shows the different types of screws and general screw construction, while Table 4-3 relates screw sizes and dimensions. The bottom portion of the chart shows the steel wire gauge measurement and the diameter of the screws.

Since drilling a pilot or starter hole is generally the best way to position screws, Table 4-4 takes the guess-work out of determining the proper drill or auger bit to use with wood screws. Besides showing the drill bit size, the chart also indicates the size number of different screws and the general body diameter of the screws.

For those projects using lag screws, Table 4-5 should prove useful. This chart shows the general availability of different sizes of lag screws and relates commonly found lengths and diameters.

BOLTS

There are two times when bolts should be used. When great strength is required or when the item being made will be taken apart frequently. Nuts are used in conjunction with bolts for fastening and washers are frequently used to protect the surface of the item being fastened. The correct bolt depends on the project being built.

Table 4-3. Screw sizes, lengths and steel wire gauge diameter.

Length (in.)	0	1	2	3	4	5	6	7	8	9	10	11	12	13	14	15	16	17	18	20	22	24	26	28	30
1/4		x	x	x																					
3/8	x	x	x	x	x	x	x	x	x	x															
1/2		x	x	x	x	x	x	x	x	x	x	x	x												
5/8		x	x	x	x	x	x	x	x	x	x	x	x	x	x										
3/4			x	x	x	x	x	x	x	x	x	x	x	x	x	x	x								
7/8			x	x	x	x	x	x	x	x	x	x	x	x	x	x	x								
1				x	x	x	x	x	x	x	x	x	x	x	x	x	x	x	x	x					
1 1/4					x	x	x	x	x	x	x	x	x	x	x	x	x	x	x	x	x	x			
1 1/2					x	x	x	x	x	x	x	x	x	x	x	x	x	x	x	x	x	x			
1 3/4						x	x	x	x	x	x	x	x	x	x	x	x	x	x	x	x	x			
2						x	x	x	x	x	x	x	x	x	x	x	x	x	x	x	x	x			
2 1/4						x	x	x	x	x	x	x	x	x	x	x	x	x	x	x	x	x			
2 1/2						x	x	x	x	x	x	x	x	x	x	x	x	x	x	x	x	x			
2 3/4							x	x	x	x	x	x	x	x	x	x	x	x	x	x	x	x			
3							x	x	x	x	x	x	x	x	x	x	x	x	x	x	x	x	x		
3 1/2									x	x	x	x	x	x	x	x	x	x	x	x	x	x	x		
4									x	x	x	x	x		x		x	x	x	x	x	x	x		
4 1/2													x		x		x		x	x	x	x	x	x	x
5													x		x		x		x	x	x	x	x	x	x
6															x		x		x	x	x	x	x	x	x

Gauge and diameter													
Steel wire gauge	17	15	14	13	12	11	10	9	8	7	6½	6	5	.135
Diameter (inches)	.054	.072	.080	.091	.105	.120	.148	.162	.177	.184	.192	.207	½	0
Steel wire gauge	4½	4	3	2½	2	1	00	00½	000	0000	—	.306	.331	—
Diameter (inches)	.216	..225	.243	.253	.262	.283	.362	.393						

Table 4-4. The correct drill and auger bit sizes to use when positioning wood screws.

Screw size No.		1	2	3	4	5	6	7	8	9	10	12	14	16	18
Nominal screw Body diameter		073 5/64	.086 3/32	.099 3/32	.112 7/64	.125 ⅛	.138 9/64	.151 5/32	.164 11/64	.177 11/64	.190 3/16	216 7/32	.242 15/64	.268 17/64	.294 19/64
Pilot hole	Drill size	5/64	3/32	7/64	7/64	⅛	9/64	5/32	11/64	3/16	3/16	7/32	¼	17/64	19/64
	Bit size								—	—	—	4	4	5	5
Starter hole	Drill size	—	1/16	1/16	5/64	5/64	3/32	7/64	7/64	⅛	⅛	9/64	5/32	3/16	13/64
	Bit size	—	—	—	—	—	—	—	—	—	—	—	—	—	4

Desired length, diameter, threads, style of head and type are the parameters that must be considered. Washers are placed between the nut and wood surface or between both the nut and head and their opposing surfaces to prevent damage to those surfaces and also to permit greater tightening ability.

Carriage bolts are used in assembling projects; they are not meant to be driven. There are three types of carriage bolts: *square (common), finned neck* and *ribbed neck*. Each has a round head. The portion of the bolt just below the head grips the material in which the bolt is positioned. This keeps it from turning as the nut is positioned on the screw thread. The finned type gets its name from the projections leading from the head to the shank. The ribbed variety has splines on all or part of the portion of the bolt just beneath the head. Carriage bolts are postioned by drilling a hole in the wood so that the shank of the bolts fits snugly into it and the top of the hole is countersunk so that the head of the bolt is beneath the surface of the wood. The bolt is then positioned using a hammer to force the bolt through the hole. The different types of carriage bolts are shown in Fig. 4-5.

Carriage bolts are generally used when wood is being fastened to another piece of wood. When they are used to fasten lumber to metal, the head of the bolt would fit into the piece of wood. Occasionally, metal surfaces will be pre-drilled to accommodate the head of a carriage bolt. This type of fastener can then be used to join metal parts. One quarter-inch to one inch sizes are typical carriage bolt diameters. They are available in three quarter-inch to 20-inch lengths. Table 4-6 is a table of carriage bolt sizes.

Table 4-5. Lag screw sizes.

Lengths (inches)	Diameters (inches)			
	¼	⅜, 7/16, ½	⅝, ¾	⅞, 1
1	x	x		
1½	x	x	x	
2, 2½, 3, 3½, etc., 7½, 8 to 10	x	x	x	x
11 to12		x	x	x
13 to16			x	x

Fig. 4-5. Different types of carriage bolts.

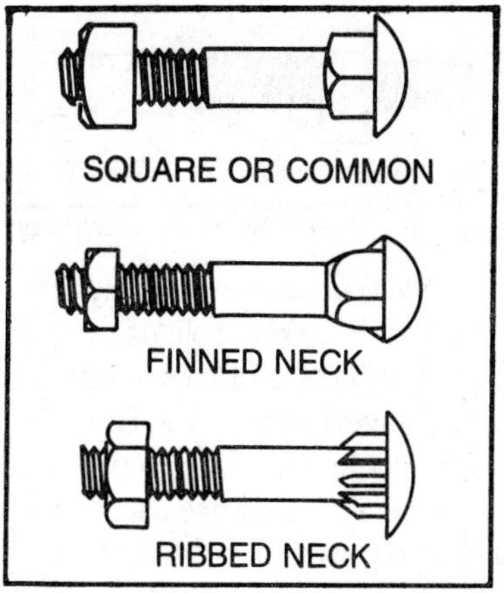

Machine bolts somewhat resemble carriage bolts, but they don't have the shoulder just below the head. They come in a variety of sizes. Those that are less than six inches long have a screw thread with a length of twice the diameter plus one-quarter inch. Longer bolts have a screw thread length that is twice the diameter of the bolt plus one-half inch. Because they are precision made, they are used for metal-to-metal applications where accurate sizing is important. As Figure 4-6 shows, they usually have square or hexagonal heads and the nuts that are generally used with them have the same shape as the bolt head. Deciding on the correct machine bolts to use for a project isn't easy. Factors such as head style, correct length, diameter, number of threads per inch and the coarseness of the thread must all be considered. Following plans or manufacturer's recommendations is best in most cases. Figure 4-8 shows the different sizes of machine bolts.

When machine bolts are used, the hole through which the bolt is to be passed is made the same diameter as the bolt itself. As Table 4-7 shows, practically any length machine bolt is available. Their diameters vary from one-quarter inch to 3 inches.

Table 4-6. Carriage bolt sizes.

Lengths (inches)	Diameters (inches)			
	3/16, ¼ 5/16, ⅜	7/16, ½	9/16, ⅝	¾
¾................................	x	......	......	......
1................................	x	x	......	......
1¼..............................	x	x	x	......
1½, 2, 2½, etc., 9½, 10 to 20.	x	x	x	x

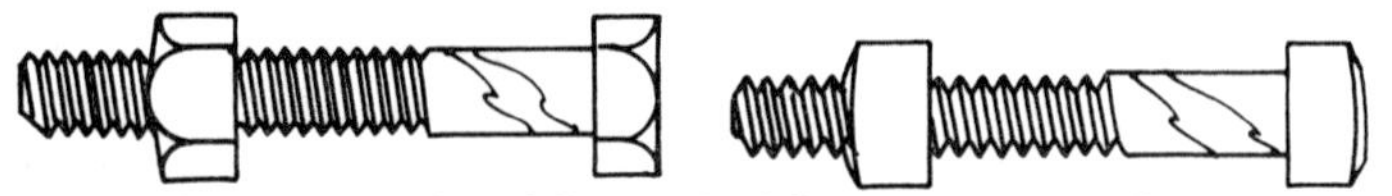

Fig. 4-6. Different types of machine bolts.

Stove bolts are probably what most carpentry craftsmen think of when the term "bolt" is used. They are not made as precisely as machine bolts and have either flat or round slotted heads. They often have heads extending over the full length of the body, but might come with threads over part of the body or covering most of it. Generally, they are used in metal-to-metal applications, but they are also commonly used to fasten wood-to-wood or wood-to-metal. When the flathead variety is used, countersinking is recommended. The roundhead type should be positioned flush to the surface. Figure 4-7 shows both types of stove bolts. Notice that the picture shows the bolts being used in conjunction with square nuts. This is the variety most often used with this type of bolt.

Expansion bolts aren't usually used in basic carpentry craft projects. They are intended for anchoring items to substances in which a threaded fastener wouldn't hold well. These bolts are really nothing more than a bolt together with an expansion shield. This shield is inserted in a pre-drilled hole and expands when the bolt is positioned within it. The shield then fits tightly within the hole and provides the fastener with a firm grip. Figure 4-8 shows how an expansion bolt would be positioned through a wooden cleat into a material that normally wouldn't hold a threaded fastener.

FASTENERS FOR GYPSUM AND PLASTER

Most carpentry projects are made of wood or in conjunction with other materials. While wood will conform to nails, screws, and other fasteners, other materials often lack this resiliency. Gypsum and plaster are such materials and they require special fasteners for proper holding.

Table 4-7. Sizes of machine bolts.

Lengths (inches)	Diameters (inches)				
	¼, ⅜	7/16	½, 9/16, ⅝	¾, ⅞, 1	1⅛, 1¼
¾	x				
1, 1¼	x	x	x		
1½, 2, 2½	x	x	x	x	
3, 3½, 4, 4½, etc., 9½, 10 to 20.	x	x	x	x	x
21 to 25			x	x	x
26 to 39				x	x

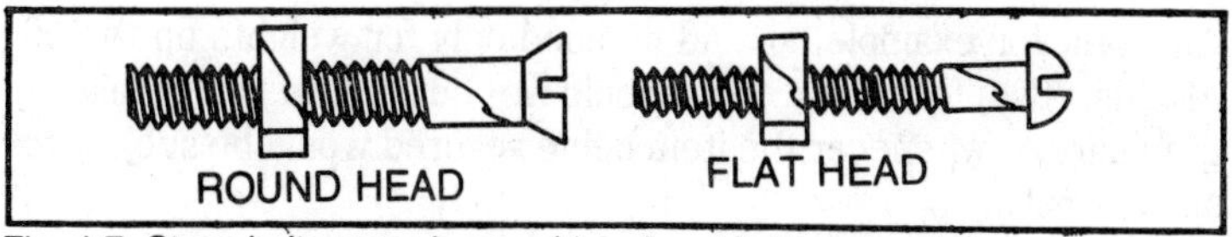

Fig. 4-7. Stove bolts come in round-head and flat-head varieties.

The secret for fastening to plaster or gypsum wallboard is to distribute weight over a greater area than a nail or screw. A spreading mechanism is provided to accomplish this.

There are two facts to remember when using these special fasteners. First, it is important to select the proper size. Expansion anchors are made with a shank that is intended to be the same size as the thickness of the wall or ceiling in which they are to be used. The second important fact to remember is that the screw used with these fasteners should not be over tightened. This is often done by individuals who aren't familiar with expansion anchors in the mistaken belief that the tighter the screw—the greater the holding ability. Actually, tightening the screw too much forces the anchor into the wall. The more the anchor cuts into the wall material—the weaker its grip.

Another device for fastening in plaster or wallboard is the *toggle bolt*. This fastener has two "wings". A hole is drilled in the plaster or wallboard and the toggle is inserted with these wings folded. As the screw or bolt is positioned and tightened, these wings spread, holding behind the wall or ceiling and allowing secure fastening.

WALL FASTENING TIPS

Although expansion anchors or toggle bolts are intended for fastening to plaster or wallboard, they do have some limitations.

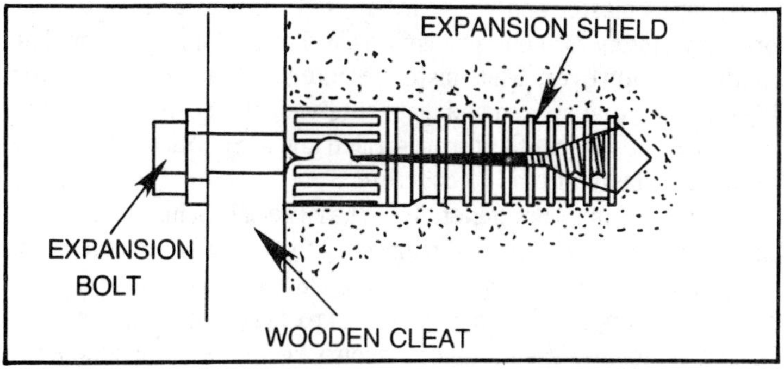

Fig. 4-8. An expansion bolt is positioned using a shield to provide a firm hold in difficult materials.

Anchors, for example, should be used only for weights up to 100 pounds. Also, for safety, they should **not** be used for safety rails or grab bars or whenever the item being secured would be subject to sudden pulls.

Whenever it is possible to do so, it is preferable to secure heavy objects or objects subject to sudden stresses directly into the wooden supports (studs) behind the wall.

Two or more anchors can sometimes be used effectively to distribute weight more effectively and over a greater area than a single toggle bolt or anchor. Use them to fasten a strip of wood to the wall. Then attach the object to the wood strip. Of course, the strip might be visible so take this into consideration.

Often, the use of plaster or gypsum fasteners can be avoided completely. For example, when hanging lightweight pictures or ornaments, a simple nail positioned at a 45° angle will disturb the wall the least. Many heavier items can be afixed with conventional picture hooks. Follow maximum weight recommendations on the package label. Damage to plaster will be minimized if a small guide hole is made for the nail.

Sometimes, you will want to reposition an object that is fastened to a wall. If only a nail was used to position the item, remove it by gently prying it from the wall with pliers. Pull in the direction of the nail shaft. If necessary, touch up the hole in the wall with a dab of spackling compound. If a toggle bolt or expansion anchor was used, try to remove the screw. Then position the anchor shield far enough into the wall so that it can be covered with a small amount of spackling compound.

MASONRY FASTENERS

Just as plaster and wallboard are not resilient enough to hold ordinary fasteners, concrete isn't as well suited as wood for that hardware. However, while fasteners intended for use with plaster have frames that spread and distribute weight, those for masonry use have sleeves that fit into pre-drilled holes. Special nails can also be used to fasten items to concrete.

Although masonry generally isn't a large part of most carpentry projects, basic techniques for masonry fastening are useful when wooden items are to be attached to brick, stone, or cement. Once some experience is gained in the use of tools for wood, fastening to masonry becomes easier. Since masonry fasteners are positioned in pre-drilled holes, the secret to good masonry fastening is the proper drilling of the holes to receive the fasteners.

Since hole-making is so important, special tools are devoted to this task. *Star drills* are used for making small, accurate holes. They are nothing more than a rod with a star-shaped tip and are struck by a hammer in order to make holes.

HOW TO USE A STAR DRILL

- Mark the exact spot where the hole is to be made.
- Hold the drill firmly and strike it with a hammer about five times.
- Loosen your grip on the drill so that it can move a small distance when struck by the hammer.
- Repeat the third step until the desired hole is completed.

Note: Better results are achieved by using many light hammer blows rather than a powerful ones. While this often involves considerable patience, a more accurate hole will result and there will be fewer cracks.

Star drills are available in three-sixteenths to 2 inch sizes. However, for holes larger than about three-quarter inch it is a good idea to drill a small pilot hole.

Electric drills are useful for fast results. However the use of special bits is required and their sizes are limited. With the proper bit, electric drills will make accurate holes cleanly.

For large holes, an *electric hammer* is recommended. It works by turning a bit and hammering at the same time. Unless you are working on a special project that requires many large holes in masonry, there is no need to purchase an electric hammer. If necessary, this tool can be rented. But, experience using tools and building projects should precede any such special project. However the masonry hole is made, the final step is to position a suitable shield and then fasten the object to the shield.

HOW TO SELECT FASTENERS FOR NON-WOOD SURFACES

There are several points to consider when working with non-wood carpentry materials. First, determine the type of project and the materials required for its completion. Then obtain the hardware for assembling the project. It is necessary to consider not only the material in which the hardware will be positioned, but the other items being joined with them as well. The hardware required to join wood to wood will generally be different than that required to fasten metal to wood or metal to metal. Be sure to consider whether the items to be fastened will be fixed permanently or if they will move as, for example, the axle on a serving cart. Strength is another important consideration. Don't try to get by with a light screw when a heavy duty bolt would be stronger.

Some materials are difficult to fasten to others unless the proper hardware is used. Even when fastening is accomplished, the results could eventually lead to problems and disappointment. For example, if two different metals are fastened together and placed in an area that has a wide temperature variation, the joined metal might bend. This happens because different metals have different rates of expansion. Even if the two metal items you want to fasten are made of the same material, problems could still arise if the fastener is made of a different metal. For example, if aluminum is used because the project will be exposed to weather, but steel screws are used for fastening—the steel may rust and ruin the job.

DON'T BE CONFUSED ABOUT GLUE CEMENTS OR SEALERS

Ever since the introduction of "miracle glues," the public has heard radio and televison commercials and seen newspaper and magazine advertisements concerning them. Supposedly, the glues can be used whenever fastening is desired and with almost any material. The unfortunate aspect of this is that we've begun to accept this premise. We've almost become "glued" on the idea.

While most of the claims for these wonder adhesives are true, they fail to mention a few very important points. First, the permanent bonding afforded by the use of these products might not be an advantage in some instances. For example, suppose the project is designed to be dismantled for easy storage. Joining components with these strong glues would not permit this, nor would their use allow parts to be adjusted at a later time. Also, glues should not be used when there are gaps that must be filled. That is the job of *caulks* and *sealants*.

Another occasion when the application of glue is not recommended by manufacturers is during severe weather conditions. Glues generally don't bond well at temperatures below 70°. When the temperature is high, the drying time of some glues is decreased. But so is their useful life.

How to Glue Properly

When using any type of glue, its best to follow the directions printed on the package. The following steps are generally required:

- Make sure that the surfaces to be joined are as clean as possible.
- Since glues are not intended to fill gaps between surfaces, make the surfaces you will join fit closely.

— Determine if the material being fastened is porous (such as wood) or non-porous (such as glass). there are different techniques for different materials.
— For porous materials, spread a thin coat of glue on each surface to be joined. As it dries it tends to be absorbed into the material and good bonding will not occur. Let this thin coat dry on each surface without joining them. Once the first coat has dried, apply a second coat.
— For non-porous materials, spread one thin coat on each surface to be joined.
— Press or clamp the pieces together. The length of time they are held together depends on the type of glue. If a recommended time is not known, clamp the pieces and leave them that way as long as possible. Over night is a good idea.

Gluing Tips

Sometimes wooden surfaces don't match exactly. An old carpenter's trick is to sand a piece of the wood from an area that won't show, or to saw a piece of similar wood to get a little sawdust. This sawdust is then mixed with the glue to form a substance that can be used to fill the gaps as well as join the items together.

Once objects have been glued together, resist the temptation to touch or move them to see if the glue has dried. Impatience will often ruin an otherwise fine gluing job.

The use of the proper glue is as important as the gluing technique. Both have a great impact on the outcome of the jobs. *Animal* or *fish glues* are the oldest of the glues designed for wood. They work well, are easy to use and, because they can be loosened with heat and water, they permit adjustment of parts and later regluing. However, this advantage also causes a restriction. Because this type of glue can be loosened in this manner they should only be used indoors where there is only a small variation of temperature and humidity.

Polyvinyl or *white glue* is a fine product for general household use. It works on porous surfaces, bonds quickly, and won't stain. It's forgiving in the sense that the joint created will have a small amount of resiliency. White glue can be used with paper or plastic items. However, it is also affected by heat and moisture.

A good general purpose outdoor product, *resorcin glue*, is somewhat expensive. However, if a strong waterproof joint is required, this product might be your best bet. Because of its strength and long lasting features, this glue is often used in boat construction and where durability is needed. It also affords the advantage of holding well on oily or resinous wood. However resorcin glue has some distinct disadvantages. It is flammable when wet, leaves a dark line when it dries and often shows through paint.

A very strong glue for non-porous items is *epoxy resin glue*. It can be used where moisture or humidity are problems and can be applied at somewhat lower temperatures than other glues; clamping is not an absolute requirement. Bonding occurs fairly quickly and complete bonding results in 12 hours to one day. Be careful when using epoxy resin glue. Although it can be used with an extremely large selection of materials, it will dissolve some plastic items.

When you can't clamp certain items, a special glue can be used. Because a bond is made as the adhesive is applied and the objects placed together, it's called *contact cement*. Generally applied with a brush or sometimes with a roller, this type of glue is usually used to bond large items such as wall panels. The most convenient type of contact cement for home use has a water base. This means that application materials can be cleaned with soap and water and used again. There is also a lacquer-based adhesive that requires lacquer thinner for cleaning up. Care should be taken when the lacquer based type is used since it is flammable.

Contact cement is unforgiving. Once the cement is applied and the pieces are together, the bond is permanent. Therefore, make sure that the parts will fit together properly **before** the glue is applied. Applying one or more coats to each piece and allowing them to dry before assembling the parts generally gives better results. Carefully follow the directions on the package.

HOW TO DETERMINE WHICH ADHESIVE IS BEST

Simply stated, the best adhesive is the one that satisfactorily accomplishes the job. There isn't a need to use contact cement if only two pieces of paper are being joined. On the opposite extreme, glue intended for use on paper won't hold a part that must bear weight. The best way to determine which adhesive to use is as follows:

— Determine how you will use the adhesive. Will it support weight, join light pieces of wood together or attach an object to an item that is exposed to weather.

— Make a list of the types of glue that can be used to satisfy the purposes in the first step.

— Consider which of the glues from the list can be used on the materials you are working with.

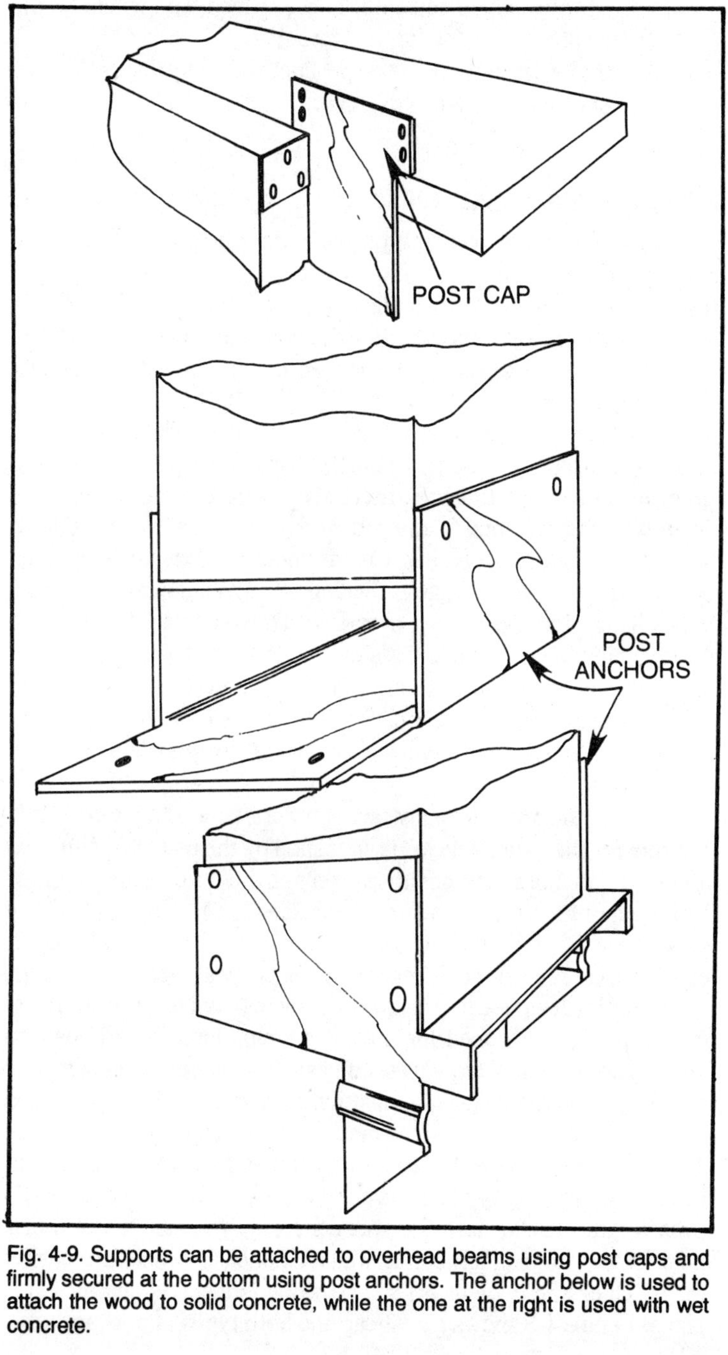

Fig. 4-9. Supports can be attached to overhead beams using post caps and firmly secured at the bottom using post anchors. The anchor below is used to attach the wood to solid concrete, while the one at the right is used with wet concrete.

— Of those glues selected, decide which glue would be the most convenient for you to use.
— If there is more than one product still on the list, select the one that is least expensive.

METAL FRAMING CONNECTORS

Sooner or later, you will probably decide to construct some sort of building item. While most of these projects will be small, those who are highly enthusiastic might feel that they are ready for large jobs that involve special tools and construction materials. Any convenience items will surely be appreciated by the individual doing this type of work.

For example, heavy duty hardware such as nails, screws, and spikes will probably be used. Using this hardware can be tedious and time consuming. Fortunately, there are special connectors that can often be used to simplify the job. These are readily available at builder's hardware stores as well as most lumber yards and can simplify framing tasks while providing strong joints. Most require only nailing to be positioned properly. They save a lot of time on construction jobs. However, it's quite important to use the proper nail with each connector. Using the wrong type of nail can lead to poor fastening or split wood. Usually, barbed nails that fit the connectors snugly and hold well in wood are supplied with the connectors.

One of the projects many carpentry craftsmen like to construct is a roof for the patio. The job is simplified by the use of *post anchors* and *caps*. Anchors hold posts securely at the bottom and the caps lock them into overhead beams. Because 4 × 4, 4 × 6, and 6 × 6 lumber are the types that are generally used for framing and support of a large number of construction projects, post anchors and caps are specially designed for this wood. As you might expect from their name, anchors are used for holding framing lumber and posts to keep them from moving. To accomplish this, one end is designed to be fastened to the building foundation. One type uses a lag bolt and expansion plug to fasten the anchor to existing concrete. The lumber is then inserted in the other end and secured with special nails. The second type of anchor is used with wet cement. By pressing the anchor into the wet concrete, you can create a firm hold for the framing lumber as well as establish a moisture barrier for it. Once the cement dries, the wood is inserted and nailed in place. Figure 4-9 shows a post cap and both types of post anchors.

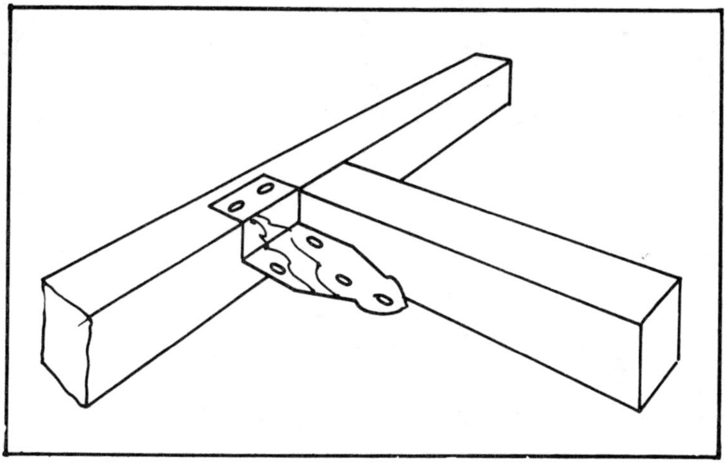

Fig. 4-10. Joist hangers are convenient to use and provide strong supports for wood framing.

Another type of framing connector is the *joint hanger*. It is used to fasten floors, decks and roof joists to beams. They are very easy to use, are strong and can be used with 2 by 4 and 2 by 6 lumber. They also come in two slightly different forms. One form is used for connecting joists to beams while the second fastens joists to flat surfaces. Both types are positioned with nails. For the carpentry craftsman who is installing a patio roof, the flat joist hanger helps in attaching the patio roof to the wooden wall. An example of a joist hanger is shown in Fig. 4-10.

Of all the devices used for connecting framing wood, perhaps the most useful is the *framing anchor*. They are useful for fastening framing lumber at angles or when two such boards meet at their middles. To allow for versatility, framing anchors are available in straight or bent types; simply choose the ones that best suit the project needs.

In order for any construction project to turn out well, it's important for the framing to be firmly positioned. Since metal framing connectors allow this to be accomplished easily and quickly, it's a good idea to become familiar with them before you begin any construction work. Knowing the best connectors to use for your projects and how to utilize them effectively will reduce the construction time, avoid errors and result in a better job. Visit builder's hardware stores or your favorite lumber yard and examine the supply of these and other aids. Sales people should be happy to answer any questions you might have and offer suggestions to help you with your specific project.

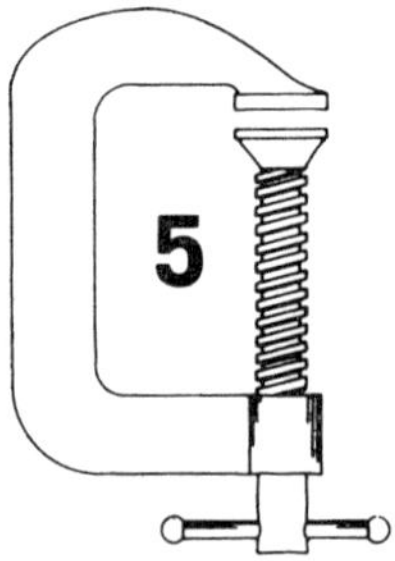

5 Special Purpose Materials

Although wood and hardware are used in every type of carpentry project, there are other materials that are used for special projects. These include products used only outside the home and some used inside a home or apartment. There are others that are intended for use within walls. These include caulks, sealants, preservatives, insulation and flashing.

FLASHING

One of the major problems with outside carpentry projects is their exposure to weather. Even a small amount of water in an area that should be dry can ruin a professional-looking job and result in considerable repair expense. One special material for combating weather is *flashing*. Made of strips of watertight, rustproof material such as aluminum galvanized sheet metal, copper alloy, or asphalt, flashing is designed to deflect water away from places that might be susceptible to water leakage. Some of these places include the lines along where roof surfaces intersect, the region where roof shingles or surfaces and dormers meet, around chimneys and next to skylights.

Another form of flashing is called valley flashing. Valley flashing provides extra protection for roof areas that serve as water channels. Valleys are designed to guide water as it drains from slopes. Otherwise, the large volume of water would eventually wear out the surface and develop leaks to the structure below. Because water flows along valley flashing, nails should only be driven along the edges of the flashing. Sometimes valley flashing has a crimp on the sides. Any puncture of the region more than an inch inside the crimps will—in almost all cases, cause a leak. Ask your roofing material supplier for detailed instructions on how to install valleys. Instructions can be found on the back side of an asphalt roof shingle bundle.

ANYONE CAN USE CAULKS AND SEALANTS

Wood isn't perfect. This imperfection can provide much warmth and beauty. However, it also provides a problem. Wood is rarely straight along its entire length. When two somewhat long wooden items are placed next to each other or when such wood is placed next to perfectly straight materials such as metal door or window frames, there is often a small gap between the items. The most effective way to fill these gaps is through the use of *caulks*. Somewhat resembling a thick toothpaste, caulks come in a cylindrical container with a tapered tip. This tip permits the carpenter to select the width of the caulking strip. All that is required is to use a scissor to cut the tapered tip at the point of the desired width.

Caulks are available that bond to many different materials and they come in a wide variety of colors. If the available colors don't match the surrounding wood, they can usually be painted.

Caulks are easy to apply. Although the tubes in which they come can be squeezed by hand, for large applications the use of a *caulk gun* is recommended. Caulk guns are inexpensive and consist of half a cylinder and a grip to hold the caulk container. They have an adjustable spring and trigger. As the trigger is squeezed, the spring presses against the caulk tube forcing the caulk from its container. Use of a caulk gun speeds the application time and one-handed operation makes caulking in difficult to reach areas considerably easier.

Don't be afraid to apply caulk if you have no experience. Caulk tends to shrink as it drys and ages, so if you apply too much the job will look better after it has set. Applying too little caulk won't hurt anything as long as there is a sufficient amount to fill gaps and bond to adjoining surfaces. However, reapplication might be required sooner than anticipated due to shrinking.

Caulking Tips

When using caulk, particularly with a caulk gun, try to get a smooth application. The produces a better looking job and makes painting easier if you are planning to match adjoining colors.

Don't plan on caulking when the temperature is very high or when it is cold. The caulk might not adhere properly. Directions on the package generally detail the correct application temperatures.

Sealants

It's probably obvious to most people that wood and water don't mix. What is less obvious is that water vapor and severe humidity also take their toll through warping and eventually rotting unprotected wood that is used outdoors. Damage to the finish and grain of inside wood can also occur. To guard against this damage, there is a wide variety of sealants available.

If you are strictly concerned with water protection, the best sealant to use is one of the resin based products. They are applied by soaking the wood in a tub filled with the clear resin sealer or by applying the sealant with a brush. The wood should be allowed to soak for 5 to 6 minutes. Soaking the wood is the preferred method, since more sealant is absorbed into the wood than by brush application. If brushing is the only method that is practical for you, be sure to apply the sealant in large amounts. With this method, clear resin sealant isn't absorbed too deeply into the wood. However, the natural color of the wood is preserved and the wood can be painted.

While resin sealants are fine for water protection, they don't give complete outdoor protection. They do protect against moisture; resin sealants don't guard against other wood enemies such as insects and fungi. Additional protection against decay is required. Areas of particular vulnerability are where wood is positioned in the ground, where there is considerable moisture and poor air circulation and around joints and supports.

To combat these enemies of wood beauty and longevity two main types of *preservatives* are used. The first type is oil based such as *creosote*, *pentachlorophenol* or *copper naphthenate*.

One preservative that has been used for many years, creosote has a long life span and is particularly recommended when wood is positioned in the ground. Creosote has long been used to protect telephone poles, railroad ties and piers. It is a favorite with gardeners who apply it to decorative lumber and fence posts. However, it has a characteristically distinctive odor and is difficult to paint over since staining often occurs in those areas that have been treated. It is applied commercially under pressure, but for most individuals, soaking is the best method. Brushing also works but wood will not absorb as much of it through this technique and so it is less effective. The key here isn't to be as artistic in applying creosote as you might be with paint or stain. The idea is to protect the wood—not make it beautiful, so use plenty of the preservative.

When you need a preservative that is clean and has no odor you should consider a clear oil. It is available in two forms. One has pentachlorophenol dissolved in oil. The other contains resins as well to help make the wood water resistant. The second type is particularly useful when the wood will touch the ground. As wood ages it tends to dry. However, as the oil base in this product evaporates it leaves a coating that helps to control the aging process. The best way to apply pentachlorophenol is under pressure but brushing also works effectively. Be careful: there are some precautions that should be taken. Since the product is also used for controlling weeds in landscaping, avoid applying it if there is the possibility of spilling it on plants. This product kills broad-leafed plants. It's also a very strong chemical. Care must be taken to avoid contact with your skin. Chemicals such as pentchlorophenol should be kept away from mouth and eyes and should not be used where children can come in contact with it during the application or drying process.

Copper naphthenate is a very effective preservative that is easy to apply. A big advantage of this product over pentachlorophenol is that it won't harm plants if it is accidentally spilled. You can easily tell if copper naphthenate has been applied to any wood you purchase. It leaves a very noticeable green shade. Some people favor this shading, but if you don't it can easily be painted over. A word of caution—copper naphthenate is soluble in water will leach if it is applied over an existing coat of paint or varnish. Its use on painted or varnished wood should be restricted. Copper naphthenate is more expensive than pentachlorophenol—a factor which should be considered if large amounts of wood are to be protected.

Another type of preservative you can apply to protect wood is water-based salts. They are known by such chemical names as *acid copper chromate, chromated zinc arsenate, ammoniacal copper arsenate* and others. They are used when you require clean wood that does not have an odor and can be painted. Besides being difficult to pronounce, this type of preservative has some real drawbacks. In order to be truly effective, each of these items must be applied under pressure rather than brushed on. Once applied, they tend to leach more readily than the oil-based varieties. And, they are poisonous to plants and shrubs.

No matter which type of preservative you selected, the important point is to choose one that meets the specific needs of the project while at the same time being easy to apply, paint or finish at a reasonable cost. But no matter which one you select, the idea is to protect the wood fom damage and decay.

INSULATING MATERIALS

Most people are aware of the drafts and discomfort that is present in homes that are insulated. We are all becoming increasingly aware of the high cost of energy for heating and cooling houses without suffcient insulation. However, most people don't know that insulation also protects the home. Whenever warm air inside a house is permitted to escape into the cooler outside air it condenses. The moisture that results affects the framing of the house and the outside paint. If the conditions are bad enough, blistering of the paint might even result. In fact, what appears to be a pesky leak might not be a leak at all, but the result of air leak condensation.

Because there are actually two problems associated with insulating, temperature and moisture control, the proper insulating material to use depends on the goal to be achieved. All insulation works by dividing large spaces into tiny air pockets. But some insulation does more. It provides a *moisture barrier* that stops water vapor. Usually this is accomplished with such materials as foil, polyethylene or a special asphalt treated paper called *kraft paper*. These materials can be used separately, but many types of insulation are foil or kraft faced. This avoids the trouble of doing two jobs instead of one.

There are three main types of insulation available, *blankets, fill* and *insulation boards*. The last type are rigid boards made of fiberglass. Although they are sometimes used over cement or cinderblock basement walls, they are generally used when constructing new buildings. Sometimes they are finished in vinyl. They provide effective insulation and are easy to work with. You can place firring strips over the wall and nail or staple the insulation boards to them.

The most common type of insulation to apply when wall supports are exposed is the blanket. They are made of fiberglass and are unfaced for general usage, kraft faced for moisture protection or foil faced for an extra moisture barrier. Once you are familiar with nailing or stapling techniques it is easy to apply fiberglass blankets. They come in widths to fit the standard spacing between wall studs and are available in various thicknesses. Each blanket is judged according to a resistance, or R value. The higher the R value the better the insulating properties. The exact rating to use depends

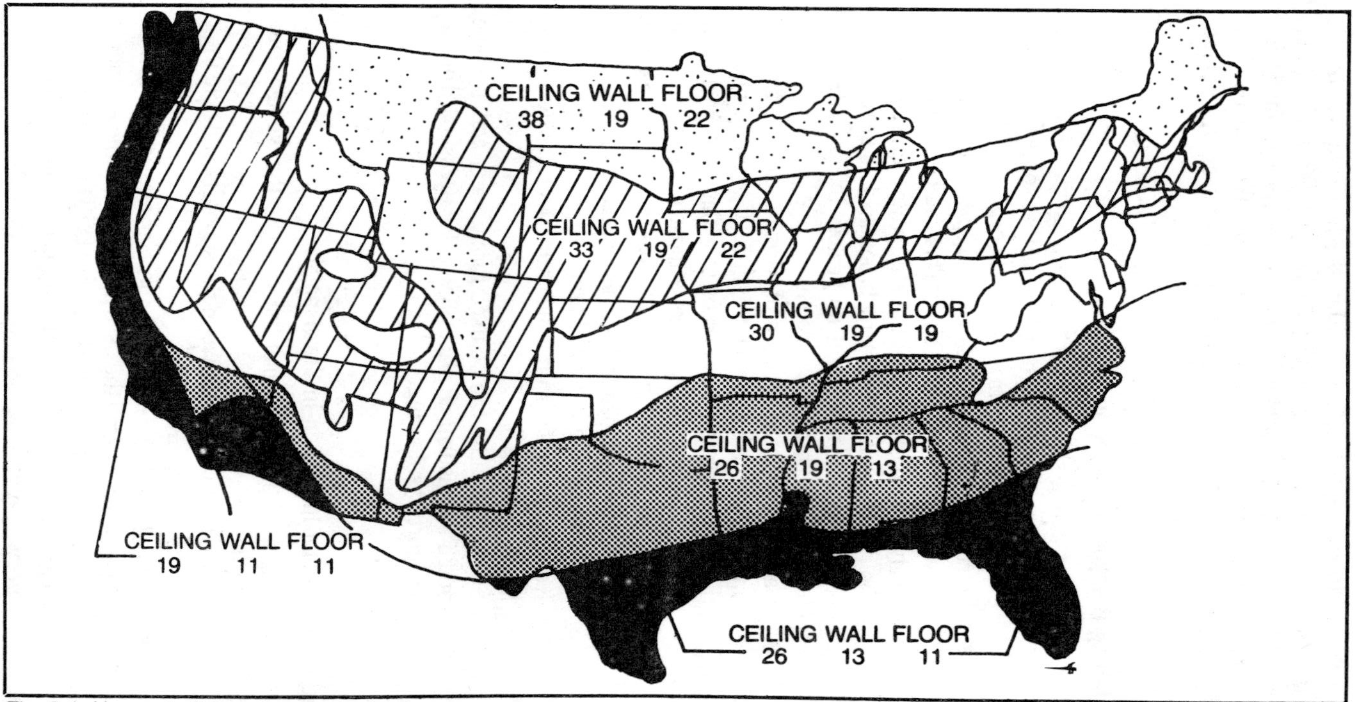

Fig. 5-1. Homes in the northen part of the country require thicker insulation.

upon the location of the insulation within the house and also the geographic region where you live (Fig. 5-1). However, since warm air rises, most heat lost in homes is through the ceilings and attic. Use thicker or higher rated insulation in the attic. When using kraft or foil faced varieties, the facing should be placed toward the heated interior area.

Installing Blanket Insulation In Attics

- Wear a long-sleeved shirt, pants and light gloves. This will protect you from the irritation that some individuals experience when working with fiberglass.
- Starting at the edges of the attic, place the insulation between the wooden beams. This can be acomplished easily from the center of the attic where there is more room or movement if a long broomstick is used. Don't place the insulation right up against the outside wall; leave some "breathing" room.
- Carefully note the location of wiring boxes for any fixtures in the ceiling below the attic. Cut the blanket so that the insulation only comes to within a few inches of the fixture box. This is done most easily before the blanket is actually positioned. **Never** cover these fixture boxes with insulation. This can cause them to overheat and results in a fire hazard.
- Repeat this process until the entire floor has been insulated.
- So that the attic can be used for storage, and as some additional insulation, place plywood or pressboard on the supports as flooring.

Installing Blanket Insulation in Crawlspaces

- Tap nails about one-half the way into the wooden supports.
- Place the blankets between the supports and hold them there with a shoulder.
- Use chicken wire to hold the blankets in position by wrapping it around the nails in much the same manner as lacing a pair of ice skates or boots. The wire does not have to be particularly strong since the insulation is not very heavy.

Installing Blanket Insulation in Unfinished Walls

- Measure the height of the wall.
- Cut the blankets into sections slightly longer than this measurement.
- Starting at the top of the wall, position a section so that it fits snugly between the studs.
- Nail or staple the insulation to the studs. Kraft and foil faced blankets have tabs through which the nails and staples can be driven.

If the walls are partly finished or if there is some access to the area behind them, pouring insulation can be used. It comes in bags that can be readily emptied into the wall. If walls are finished an insulating foam can be used. A small hole is made in the wall and the material is blown in with special equipment. Generally, this is done commercially.

There are other insulating materials that you can install yourself. Most people don't even realize how much they can increase insulation in their homes. For example, the installation of wood paneling over an existing wall cuts drafts and heat loss through the wall. If desired, insulation can also be added to provide additional protection. The installation of storm windows and doors is simple for the craftsman who is familiar with the use of tools. Since heat loss through these walls openings is a serious problem to the average homeowner, the carpentry craftsman can do much in this area to keep himself comfortable and save money on fuel bills as well.

Often, houses, even those that are well insulated when constructed, develop drafts around the windows. This is because the insulation blankets that were installed when the house was built have settled due to their own weight. The result is uninsulated areas and heat loss. You can easily correct this problem by the method listed below:

HOW TO CORRECT DRAFTS AROUND WINDOWS

- Obtain a blanket of unfaced insulation.
- Using a nail puller of hammer and sturdy screwdriver, carefully pry the molding at the bottom of the sill away from the wall. It is only necessary to make a small gap. Do not remove the molding.
- Cut or tear the insulation into pieces about 2 inches long.
- Using a putty knife, force the pieces into the gap bet-

ween the molding and the wall. Use as much insulation as possible.

- Re-position the molding. If the nails used originally to install it are convenient, tap them a few times with a hammer to position the molding firmly. If the nails can't be accessed without damaging the wood, take a scrap block, place it against the molding and strike it with the hammer instead. This should avoid damage to the molding. However, if some touch-up is required, use wood filler or pre-mixed spackling compound to fill any hammer marks. Lightly hand sand if necessary, and touch up the paint.

Other products besides actual insulation can serve as insulation. For example, caulk stops the flow of cold air into the warm interior of a house. Similarly, other materials not normally associated with insulation help the insulation effort. When paneling is installed over firring strips an air pocket is formed between the original wall and the new paneled one. This air cushion helps to insulate, while the wood itself helps stop air leaks.

Other wall coverings can be applied that not only look good but make a room seem warmer as well. Burlap and cloth wall coverings act much the same way a blanket does when you sleep. They are especially useful for decorating a normally cold wall and helping to keep the room warm at the same time. However, the warming affect will not be as great as that provided by the use of other types of in-wall insulaton.

INSULATING TIPS

When framing an addition to a house, plan the types of insulation you will use. It's easier to install insulation during the construction process than afterwards when poured or blow-in insulation must be used or the walls opened.

When using insulation, plan for the settling of the material that will occur within several months after the original insulation effort. Provide a means for adding small amounts of insulation after this settling occurs.

Cutting fiberglass blankets is easy using just a sharp knife and a straight-edge. Simply place the blanket face up on a hard surface such as scrap plywood, wallboard or work surface that has been protected against knife marks. Compress the blanket with the straight-edge and cut it with the knife (see Fig. 5-2).

Faced insulation requires some additional care during installation. To provide a proper, effective moisture barrier, the facing

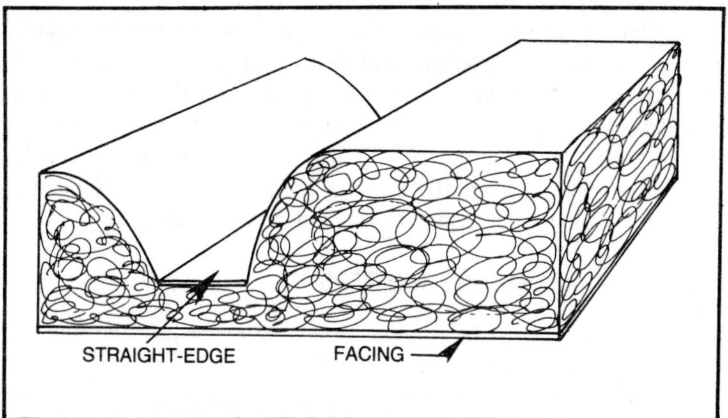

Fig. 5-2. A straight edge and sharp knife are the only tools required for cutting fiberglas blanket insulation.

must be intact. Should rips or tears occur, repair them with tape or place another piece of facing over the damaged area. It is important to make sure that the facing of the oil does not touch the back of the wall or paneling. The greatest protection is obtained by allowing at least three-quarter of an inch of air space.

If unfaced insulation was previously installed and vapor protection is desired, you can easily correct the situation. A separate vapor barrier can be installed using just a hammer and nails or a staple gun. *Polythylene* is a plastic that is sold in rolls and is available in various thicknesses. The four mil thickness is recommended. Simply unroll the film across the wall and nail or staple it to the supports, studs and framing. Placing paneling or wallboard over the vapor barrier allows the wall to be finished in an attractive manner and affords the important advantages of securing the barrier and supplying an excellent vapor seal. The polyethylene can also be used to minimize clean up if the wall will be finished. Allow the polyethylene to cover any windows until the ceiling and walls have been painted.

Unfaced insulation is recommended for attics. If faced insulation is to be used, it is recommended that the facing be removed or slashed at regular intervals. The idea is for vapor to flow freely into the attic and out the ventilators. This means that the ventilators should never be blocked by insulation or by any objects being stored in the attic.

Because of the importance of saving energy and the opportunity to save money in heating and air conditioning bills, you are well advised to plan on insulating any additions to a home.

One of the major areas of heat loss, even in a well insulated home, is through windows and doors. However, you can help overcome these losses. You can plan and build wooden storm windows that form a pocket of insulation air between the regular window and the storm window. If storm winows already are installed, attractive inside shutters can be designed that are both decorative, functional, and reduce drafts and heat loss as well.

HOW TO MAKE INTERIOR SHUTTERS

- Measure the window frame and decide how the shutters will be attached to it.
- Get some large sheets of construction paper from an art or stationary store or make your own by taping several pieces of regular paper together.
- Draw a pattern on the wood to comprise the shutters.
- Use a jig saw or sabre saw to cut the pattern in the wood.
- Sand the wood with fine sandpaper to remove any rough spots in the pattern and remove any saw marks.
- Drill any holes required for attaching the shutters to the window.
- Paint or stain the shutters and allow sufficient drying time.
- Attach the fasteners to the wood and then install the shutters on the window.

6 Carpentry Around the House

Although carpentry projects can be enjoyed in both home or apartment, there are many projects that are specifically useful to the home owner. These projects often increase the value of the home and at the same time reduce operating and maintenance expenses. Some of them involve structural work that is not visible and allows the inexperienced carpenter the opportunity to gaining valuable practice. As these projects develop, the knowledge you gain can be applied to each step until the finished result is a professional-looking job you can be justly proud of. Other jobs require some sophistication in carpentry and should be attempted only after a suitable level of familiarity with technique is obtained. The experienced craftsman will derive many hours of enjoyment from home improvements and remodeling projects. He knows full well the feeling of satisfaction of having built it himself.

HOW TO SEAL A HOUSE

One of the simplest, and most important projects for the home owner is the sealing of the house. He knows that insects can enter through small cracks in the siding or between windows and doors and the frame of the house. Drafts and heat loss results in expensive fuel bills. Water damage from leaks can be expensive to repair. Although the application of sealants isn't strictly a carpentry job, sealants don't work very well unless applied to a properly prepared surface. That's where carpentry skills play an important role.

- Examine the areas around windows and doors for cracks and air spaces. Caulks tend to shrink as it ages and even recently applied products may have some small air gaps.
- Make sure that the the surfaces are clean, dry and free of grease or oil. If necessary, wash any soiled areas and allow them to dry. If they are still dirty, lightly sand them until they are clean.
- Apply fresh caulking. If the area to be sealed is large, a caulking gun will help speed the job.
- Touch-up the repaired area with paint if necessary.
- Inspect the outside of the house for any areas that should have been caulked but for some reason never were. The procedure for caulking them is the same, but some sealants require primers to be applied before painting. Be sure to check the directions on the package.
- Check the areas of the house that are most exposed to weather. Apply a *butyl rubber* type caulk to joints in these areas. This type of caulk weathers well and resists moisture. It can be applied to dissimilar materials. However, never use any caulk on wide gaps since shrinkage will again leave gaps.
- Check the areas around skylights and basement doors. Apply a *polysulfide rubber* caulk to these areas. They are effective for use on moving joints, last for many years and resist weather extremes.
- Check the house exterior for loose molding. If nail heads are visible, a few taps with a hammer and nailset should correct the situation. If necessary, drive a few finishing nails, set them, fill the holes with a dab of wood filler, lightly sand and touch up the paint. If the nails aren't visible, place a block of scrap wood along the molding, and strike it firmly several times with a hammer. Then use finishing nails.
- Check flashing along the roof and joints. If it is damaged it must be repaired or replaced. Otherwise, expensive damage from water will result. While you are at it, be sure to examine the flashing over windows and doors. They often are overlooked during inspections. Be sure to nail flashing only along the "tabbed" portion. Using nails elsewhere will lead to leaks.
- Don't forget to examine the roof. Loose tiles or shingles can lead to damaged ceilings. Tack them down with special roofing nails.

If you've followed these steps, you've checked the four major areas where water seeps into a house (other than the basement). You examined the areas around windows and doors, and payed particular attention to the head jamb. The doors and window frames themselves were also checked, especially the area between the tracks of sliding patio doors, the sill on which they rest and any overhead jamb. The third area, where lengths of siding come together, was checked for any signs of warping or shrinkage that resulted in gaps to be filled. Finally, the joints between exterior areas, such as decks or porches, and the house itself were considered. Joists supporting decks and ceilings that form patio roofs require caulking occasionally.

WHEN TO ADD INSULATION

Once the house is sealed on the outside, the next step, is to examine it on the inside. The attic is the place where most heat is lost from the home. The attic should have 6 inches of insulation on the floor supports or heat will pass through the roof. Check crawl spaces and basement walls for insulation. Feel the walls on a cold day. If they are cold add insulation.

Check for cracks around the edges of door jambs and the tracks of sliding glass doors. If cracks are present, drafts are also. Fill the cracks with silicone tile compound. This product comes in tubes and is available at most lumber yards and hardware departments in department stores. It is waterproof, dries rapidly and is extremely easy to apply, but it doesn't provide a good surface for paint. Use a rubber based sealer if you plan to paint or stain.

Next, remove the screws holding wall plates around light switches and outlets for safety, and disconnect the electricity by removing the associated fuses or tripping the proper circuit breakers. Then fill any spaces around, but **not** inside, the junction boxes with pre-mixed spackling compound. Replace the plates, turn on the power and you have insulated these frequently overlooked spots.

Once the house is sealed and insulated you are ready to make any changes or additions. Often, one of the first modifications to any home is the addition of a wall.

HOW TO CONSTRUCT A WALL

Although a wall might look solid when it's finished, it actually consists of a large number of separate items. Each of these components has a special purpose and together they provide a solid

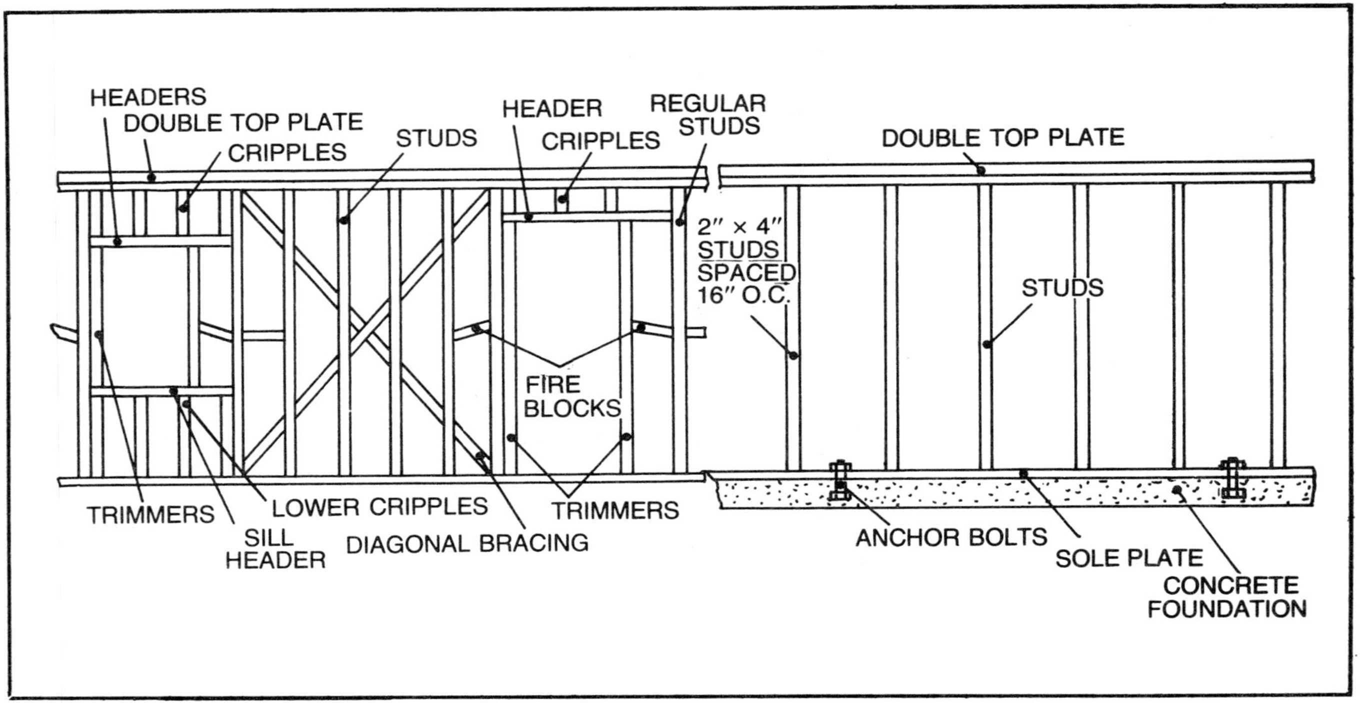

Fig. 6-1. Details of a typical wall frame.

framework over which a finishing material can be applied. Figure 6-1 shows these components and their relationship to one another. *Sole plates* are support lumber that is anchored to a concrete foundation, for new major walls, or nailed to existing supports when the proposed wall does not have to support any weight. Running vertically, other wood members called *studs* are attached to the sole plates. These studs are extremely important. They support *top plates* and the weight of anything above the top plate line. They also provide the framework to hold sheathing, for outside coverings and siding, and the lath, plaster, insulation and plaster board that is found on the inside portion of exterior walls and on inside walls. In fact, the amount of weight a wall can hold is mostly determined by the strength of the studs. There is also wood that parallels the sole plate, referred to as *top plates* because of their location at the top of the studs. Whenever a wall must support weight or to be used as a partition, it should have two top plates. Other parts of a wall include *diagonal bracing, cripples, trimmers, headers*, and *fire blocks*.

Before I begin to explain how to piece all these parts together to form a unified wall, it's first necessary to mention some important specialized items. *Corner posts* are special studs used at the corners of frame construction and usually consist of three or more ordinary

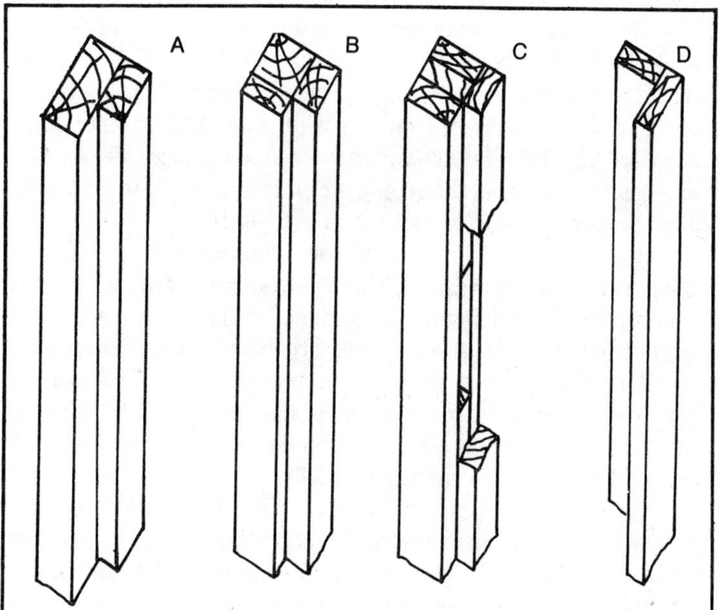

Fig. 6-2. Different ways to construct corner posts.

studs placed together for strength. At the start of construction of any weight supporting wall used as part of house framing, corner posts are constructed, plumbed so they are vertical and temporarily braced so they will stay in position while other framing components are added. The first picture in Fig. 6-2 shows a corner post for a 4 inch wall. It consists of a 4 × 6 with a 2 × 4 nailed to it so that they are flush along one edge. If a thicker wall is to be built, heavier wood must be used.

A second type of corner brace (Fig. 6-2B) consists of a 4 by 4 with a 2 by 4 nailed to two adjoining sides. Figure 6-2C shows how a corner post can be constructed with two 2 by 4s nailed together with blocks between and a 2 × 4 flush with one edge. Figure 6-2D shows a corner post that is often used when the wall will not be finished. It's made of a 2 × 4 nailed to the edge of another 2 × 4 so that the edge of one is flush with the side of the other.

When a house is remodeled, it's often necessary for a partition to join an outside wall. When this happens, a stud wide enough to extend beyond the partition on both sides is required to provide a solid nailing base for the inside wall finish. Called a *T-post*, this type of stud is made in four ways as shown in Fig. 6-3A.

The first T-post shown consists of a 2 × 4 that is nailed and centered on the face side of a 4×6. Sometimes a 2×4 can be nailed and centered on two 2×4's that are nailed together, as indicated in Fig. 6-3B. Another frequently used form of T-post is shown in Fig. 6-3C. It is made of two 2×4's nailed together with a block between them and a 2×4 centered on the wide side. Still another form consists of a 2×4 that is nailed and centered on the face side of a 2×6 and includes a horizontal bridging nailed behind them to lend both support and stiffness. It is shown in Fig. 6-3D.

Whenever a building or an addition is constructed, the first step is to provide a solid foundation. This is the purpose of the *sill* and it is the first part of a building to be positioned. Sills can be placed on a cement base or on the ground. They are joined at the corners and, depending on the size of the addition, can be spliced if necessary. Figure 6-4 shows the different types of sills that are used. Although you would probably only work with the lighter types of sills, I will describe the four different types of sills.

The first two illustrations detail *box sills*. This is the type most frequently used with platform framing and can be constructed either with or without the *sill plate*. The sill plate lies on the foundation wall or ground. The sill itself is positioned on the outside edge of the sill plate.

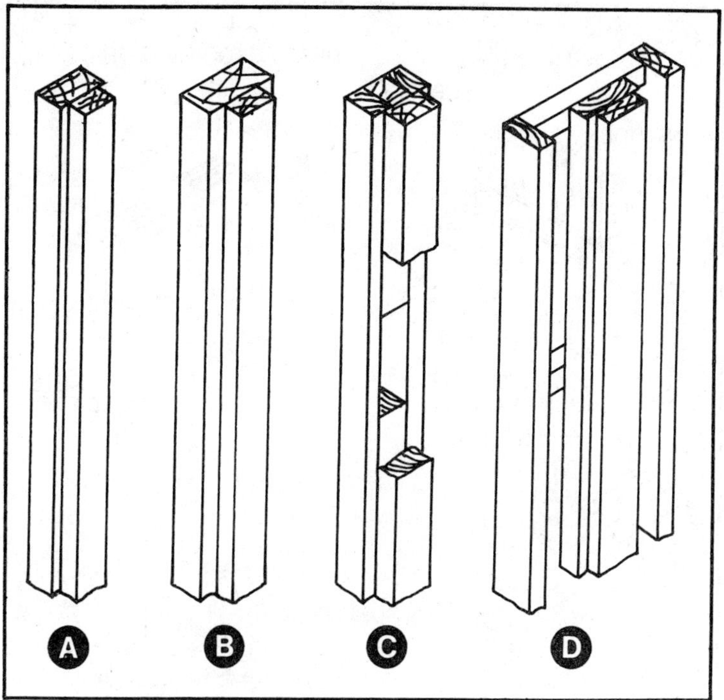

Fig. 6-3. How T-posts are constructed.

The *T-sill* has two varieties. The first T-sill, shown in Fig. 6-3C is intended for use in dry, warm climates while the variety shown in Fig. 6-4D is used in those climates that aren't as warm. Although they might not look similar in the illustration, they really are. The T-sill shown in Fig. 6-4D shows joists nailed directly to the studs as well as the sills. Headers are used between the floor joists.

Figure 6-4E shows a *braced framing sill* that is used in braced-framing construction. Note that the floor joists are notched and nailed directly to the sill and studs.

Constructing sills is not difficult, but it does require familiarity with tools. Since the entire project depends on how firmly the sills are made, it's best to study the pictures carefully before attempting to make them. Since actual dimensions will vary with the planned addition and the types of walls to be built, dimensions are not shown. If only a simple wall is to be built and the wall won't support any building weight, the foundation becomes even simpler than those shown. All that is required under these circumstances is a firm footing to build on. This can be as simple as the use of only a sole plate as show in Fig. 6-1.

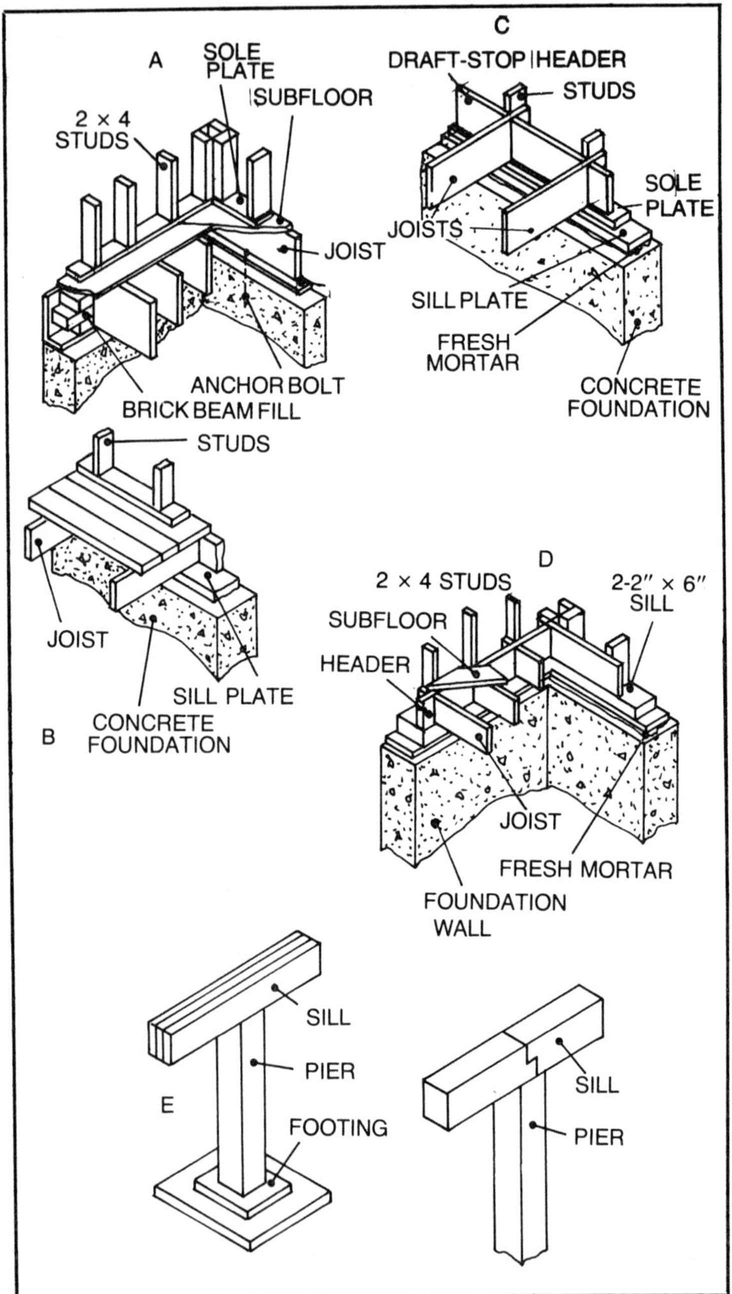

Fig. 6-4. Different types of sills. A and B are box sills. C and D detail T-sills, while E depicts a braced framing sill.

When building a simple non-supportive wall, the next step is to position a *top plate*. This serves somewhat the same purpose as the sole plate at the bottom of the wall. It provides the anchorage for the top part of the wall. For additional strength, the top plate is usually made of two pieces of framing lumber.

In those cases where the wall will support building weight, the top plate also serves as a support for the rafters. When the studs at the end of a building extend to the rafters, no plate is used at the end of the building. When it is used on the top portion of the wall, the top plate is also called a *cap*. The bottom section of a double plate is nailed to the top of the corner posts and to the studs with 16 or 20-penny nails. Figure 6-5 shows how this connection is made at the corner.

When building a wall that doesn't contain windows or doors, the next step is to position the studs. Using 16 or 20-penny nails, fasten the studs to the top plate. Begin by measuring from one corner and mark the location of the studs. They should be spaced 12, 16 or 24 inches apart, depending on the type of finish that will be used. Generally, 16-inch spacing should be used for most purposes. When vertical siding is applied, use a wider spacing since horizontal wood positioned between the studs provides the nailing surface needed for that type of wall finish.

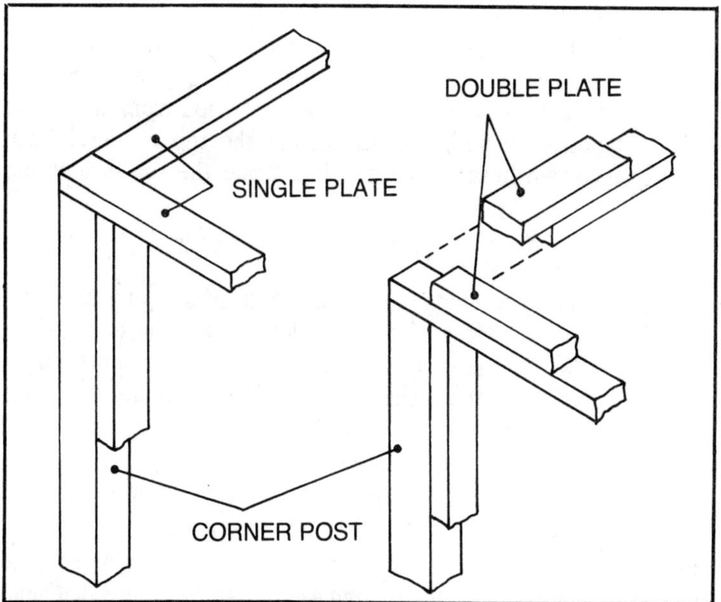

Fig. 6-5. Single and double plate construction.

For an added effect, you could break up a solid wall with a window, a service opening or a door. For example, a solid wall between a kitchen and dining room can have an opening resembling a window through which the host or hostess can pass food or beverages. In such cases, the frame for the opening should be layed out before the studs are fastened through the top plate. This is done as follows:

— Cut the stud lumber so that its length will be the height of the bottom of the window less the thickness of the sole plate and the bottom of the window itself.
— Mark the width of the window on the sole plate and nail these short studs to it at the markings.
— The wood forming the bottom of the window frame is called the *bottom header*. It might be made of one piece of framing wood or doubled. To make the header, cut framing lumber so that its length is the same as the distance between the short studs plus the thickness of *both* of the studs.
— Nail the bottom header to the short studs. If the bottom header is to be doubled, fasten one board to the studs and then nail the second to it. The ends of the bottom header should be flush with the outside surface of each short stud.
— The top of the window frame consists of a single or double *top header*. It should be the same length as the bottom header. Using the same type of wood from which the bottom header was made, cut the top header. If the bottom header was doubled, make the top header the same way.
— Cut two studs so that their length will be the desired height of the window.
— Nail these studs to the bottom header taking care that their outside surfaces are aligned with the ends of the bottom header and outside surfaces of the short studs.
— Fasten the top header to the studs positioned and nailed to these studs the same way the bottom headers were nailed to the short studs (step four).
— Place a stud against each side of the window framing and nail them to the top plate and sole plate.

Figure 6-6 shows how a window frame looks when completed. The illustration shows that constructing a door frame is somewhat similar.

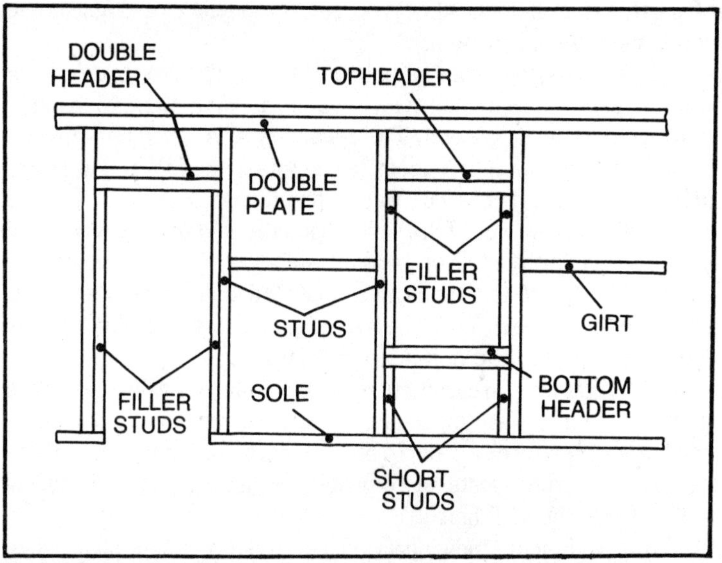

Fig. 6-6. Door and window framing showing headers and plates. Girts are used in quick construction when vertical siding will be used to cover the outside walls.

- Measure the desired door opening and mark this distance on the sole plate.
- From each of these markings, measure the distance that is the same as the thickness of a stud.
- Place studs against the outside between the sole and top plates.
- Cut two studs so that their lengths are equal to the height of the desired door opening less the thickness of the sole plate.
- Nail the two cut studs to the inside surfaces of the studs which were positioned in the third step.
- Cut stud lumber into two lengths that are the same as the distance between the outside surfaces of the studs positioned in the previous step.
- Nail these lengths to the studs.

BEARING WALLS AND PARTITIONS

Generally framed as part of a building, *partition walls* divide the inside space within the building. When the outside of a structure is to be completed prior to the installation of floors, partition walls are left unframed. Although considerable carpentry experience should be acquired before you attempt to construct a building, knowledge of partitions is helpful when you are remodeling or doing home additions and enhancements.

There are two types of partition walls. The first, or *bearing* wall, is given that name because it supports weight. Bearing partitions supports ceiling joists. The second type is called *non-bearing* because the only weight it needs to support is itself. When walls are finished, there is no obvious way to distinguish bearing walls from non-bearing. However, if the wall is open there is an easy way to tell which type is present. Non-bearing walls generally have only one cap or plate. Since this type can be installed at any time after other framework has been built, beginning carpenters usually become most familiar with the non-bearing partition.

Partition walls are framed in the same manner as outside walls and door openings are made in the same way as outside openings. Whenever there are corners to be considered or where walls are to be joined, corner posts or T-posts are used to provide nailing surfaces for the wall finish.

Although some individuals omit the use of a sole plate when installing a non-bearing wall to reduce construction time and costs, a sole plate is recommended in every case. It helps distribute the weight of the wall over a larger areas.

WALL STRAIGHTENING AND PLUMBING

Once the corner post, T-post, and intermediate wall studs have been fastened to the plates, the wall must be plumbed so that it is truly vertical. Once this is accomplished permanent braces and rafters can be placed. It is not difficult to plumbing and straightening walls but you must work carefully and obtain accurate results. There are two widely used techniques for plumbing posts and these are illustrated in Figure 6-7.

How to Plumb Posts—Technique One

- Attach a plumb bob to a string long enough to reach to, or below the bottom of the post.
- Position a rule on top of the post so that 2 inches of the rule extend over the side of the post.
- Hang the bob line over the rule so that it is at the end of it and 2 inches from the post top. Then let the bob drop so that it reaches the bottom of the post.
- Use another rule and measure the distance from the post to the bob. It should be 2 inches. If it isn't, the post

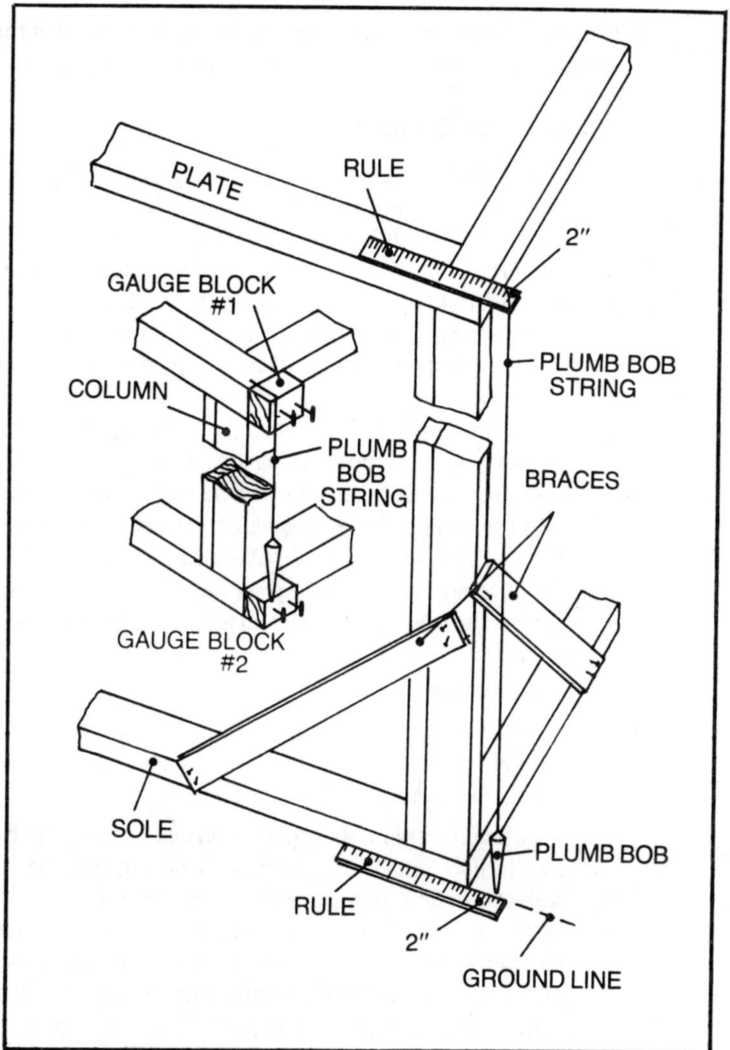

Fig. 6-7. How to plumb a post.

is not plumb. To make a correction, move the post inward or outward until the distance from the post to the bob line is exactly 2 inches. This might require several trials until the measurement is exact, but accuracy is important.

- Nail a temporary brace in place.
- Repeat the previous steps from the other outside face of the post.

The entire procedure should be performed for each post so that the wall is plumbed properly.

How to Plumb Posts—Technique Two

- Attach the bob to a string that is long enough to reach near the bottom of the post.
- Attach the plumb bob string to the top of the post.
- Obtain two blocks of wood that are identical in thickness.
- Nail the first block near the top of the post between the plumb string and the post. Don't fasten the block too tightly to the post since it will be removed when the post is plumbed.
- Hold the second block between the plumb bob string and the bottom of the post so that this block is against the post. If the entire face of the second block makes contact with the string, the post is plumb. Otherwise the post must be moved inward or outward until it is plumb.
- Nail a temporary brace in place.
- Repeat the previous steps from the other outside face of the post.

How to Straighten Walls

Whether the wall being constructed is the outside wall of a building or an inside partition wall, the same straightening method can be used. This procedure is shown in Figure 6-8.

- Use a level or a plumb bob and plumb one corner post. Be sure to use temporary braces to hold it in position.
- Repeat the first step for each corner post.
- Fasten a chalk line to the outside of one post at the top and then stretch the line to the outside of the next post. Fasten it to that post in the same position as on the first post; allow sufficient room to place small three quarter-inch blocks under each end of the line to provide clearance.
- Position temporary braces at intervals that are small enough to hold the wall straight. Don't nail them to the wall just yet.
- Take a three quarter-inch block and slide it between the line and the plate. The block should just pass between the chalk line and the wall. If it does, nail the brace to the wall. Otherwise, move the wall framing until this clearance is obtained and then fasten the temporary brace.

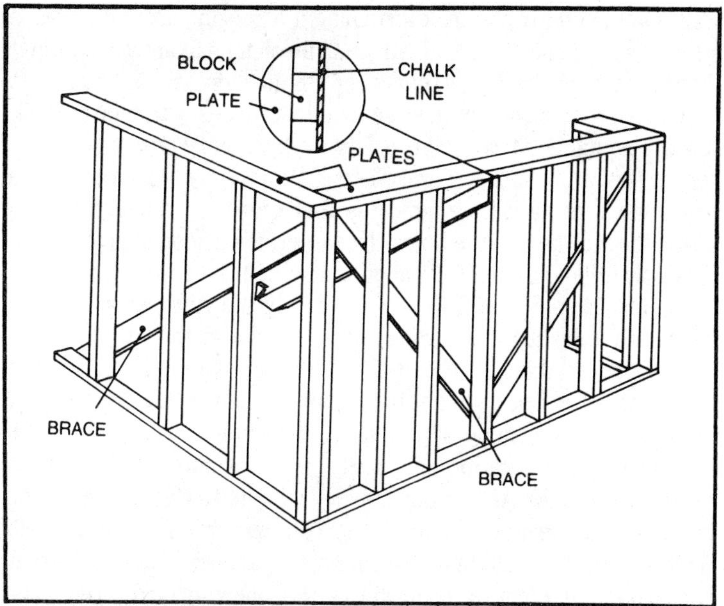

Fig. 6-8. The technique to use for straightening a wall. Proper straightening requires the use of a level or plumb bob. The insert shows the correct positioning of the chalk line.

For an outside wall, these steps would be repeated around the entire building. For an inside wall they are performed until the entire wall has been straightened and braced.

HOW TO OPEN A WALL

Often, the homeowner will decide to make an improvement that requires the removal of part of or all of a wall. This opening might be small to allow the placement of an electrical outlet or large enough to position a door. But regardless of the opening size, the procedure is basically the same. The first point to remember is that unless you are an expert in construction, major openings in bearing walls are best left to professionals. Opening a bearing wall in the wrong manner can seriously affect structural integrity. A job done incorrectly is expensive to repair.

Another point to remember is that although walls might look solid, they are generally hollow. The space inside is often cluttered with electrical wiring, plumbing or heating and air conditioning ducts. Accidentally cutting or puncturing these items not only results in expensive repairs but could lead to injury as well. The wise carpenter is always cautious whenever any opening is made in a wall. One useful trick for determining if a specific area of a wall is suitable for opening is to take a small finishing nail and *gently* tap it through the plasterboard covering. If no increase in the wall's resistance to the nail is noticed repeat the process with a longer nail. If there is still no increase in resistance try a very large nail. Then repeat this process in several places. A uniformly small resistance to the nails at each hole indicates that there is probably nothing inside the wall behind the holes. However, it's a good idea to double check by drilling a small hole about one-half inch in the region of the nail holes. Being careful to penetrate only the surface of the wall. Then use a flashlight to examine the interior of the wall.

Not everything inside a wall is "foreign". Some items such as studs, girts and braces are actually part of the wall itself. However, because they are generally connected to other wall components with nails, their location can easily be determined without even making a nail hole. An inexpensive device called a *stud finder* is used for this. An inexpensive magnetic compass works just as well. Simply move the stud finder or magnetic compass close to the wall and around the entire area where the opening is desired. The pointer will make a sharp deflection whenever a stud is encountered. Once the anatomy of the wall has been determined, opening it is a simple procedure:

- Mark the boundary of the desired opening on the wall.
- Mark a small square next to one of the corners of the boundary. Be sure that this small square is inside the boundary marked in the first step.
- Carefully drill a small hole at the corner of the square marked in the second step. Make sure that the drill just passes through the outer surface of the wall. Then place a piece of tape in the center of the square so that one half of the tape sticks to the wall and the other half hangs freely. This tape will provide a "handle" so that the little square can be removed more easily.
- Use a keyhole saw or other small handsaw and cut out the small square marked in the second step. Try to keep the saw blade from penetrating too deeply into the wall so that if there is any plumbing or electrical wiring behind the opening it won't be damaged.
- Once the square has been cut, gently work it loose from the wall by using the tape handle.
- Now that a little opening has been made, take a flashlight and look inside the wall to double check that there are no

unexpected items; if there are plumbing or wiring or ducts be sure to note their location so they may be avoided when the larger opening is cut.

— Use either the handsaw or, for more speed, a hand power saw to cut the larger opening. If it isn't too large, a tape handle will help in the removal of this piece, too. (Note: Checking How to Cut Square Corners will help you to make correct squares.)

Once the "skin" of the wall has been removed, the entire interior of the wall is exposed. If any wooden members are to be cut, be sure to brace them before they are sawed. Support vertical members with horizontal framing lumber and horizontal members with vertical structures.

INSTALLING DOORS

Installing door frames isn't as complicated as it looks. Like most carpentry projects, the secret of beautiful doors is in the framing. If care is taken when you are constructing the frame, the actual hanging of the door is simple. Problems will arise, however, if the framing is not strong, square and plumbed correctly. The actual framing construction depends on the door usage. Outside finishes and materials are required for outside doors. Climatic considerations are important here as well. Since the interior of a house doesn't have the wide temperature variations that the exterior is subject to, and moisture is more controlled, the height of an outside door and its construction will differ from that of an exterior door. Door openings in exterior walls usually require headers. Regular studs are placed 16 inches apart. Both inside and outside doors have extra studs at the sides of the door opening itself. Because the way the door is mounted will affect its operation, plumbing is an important part of door installation; one-half inch should be left between the back of the jambs and framing to permit accurate plumbing and leveling.

Exterior door framing should be completed before covering is placed on the outside walls. This is done by squaring any uneven *sheathing*. Waterproof covering is placed over walls to protect them from moisture. Heavy building paper is then wrapped over the sides and top. No paper is placed on the floor since the sill has to be positioned to a portion of the rough flooring. The paper must be adjusted so that it is even with the inside part of a stud and extends to a point that is tacked into place.

There are several different ways to construct outside door frames. The best way depends on a variety of factors such as climate, desired strength and type of door. The method described here is relatively easy and can be accomplished quickly. It requires no special frame construction because it uses the studs on each side of the opening as a frame. Figure 6-9 illustrates the construction of this type of exterior door housing.

How to Construct an Exterior Frameless Door Frame

- Determine where you will hang the door. Make sure the opening is set back the width of a stud.
- Apply outside finish to the wall. Leave the opening free of material.
- Nail the door casing to the sides of the opening.
- Nail a piece of wood threequarters of an inch by three quarters of an inch over the door. This serves as a support for the drip cap. It should be set back the width of a stud.
- Mark the casing to show the location of door hinges.
- Nail hinge blocks to the casing where it is marked.

Once these steps are completed, the door is ready for hanging.

How to Construct an Inside Frameless Door Frame

As in the case of outside door frames, inside frames can be made in several different ways. One particularly convenient method is very similar to that just described for outside use. However, an inside frame doesn't require a door casing and hinge blocks are nailed to the inside wall finish once the location of the hinges is determined. This provides a nailing surface for the hinge so that it is flush with the door. Figure 6-10 shows how a home-made slat door, which can easily be made by a carpentry craftsman, would be hung.

DOOR JAMBS

The linings of the framing of door openings are called *door jambs* and they are present in both exterior and interior doors. Figure 6-11 shows door jambs and trim and illustrates how casings and stops are nailed to the door jambs. It is from these casings and stops that the door is hung. Jambs for inside doors are made from three-quarter inch wood, while outside jambs are of one and three eighths-inch wood. The width depends on the thickness of the walls. The jams of inside doors are built up with three-eighths by

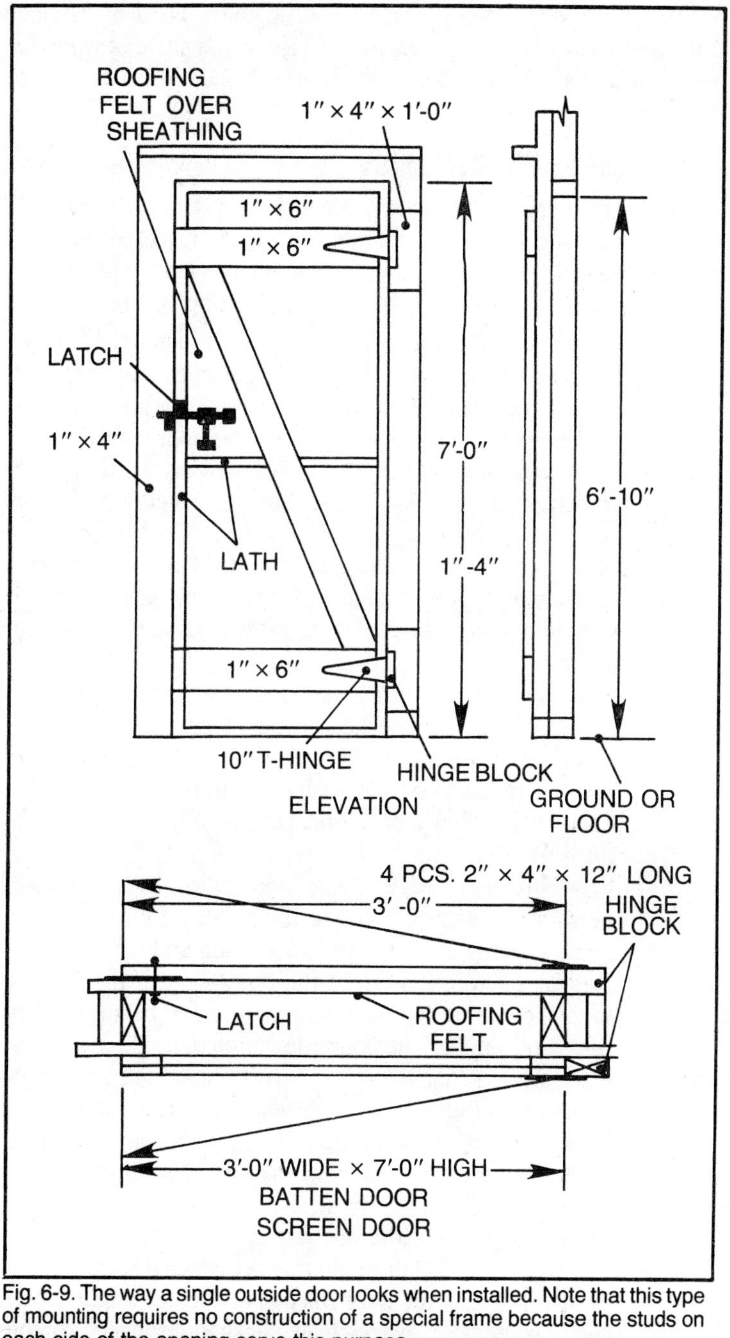

Fig. 6-9. The way a single outside door looks when installed. Note that this type of mounting requires no construction of a special frame because the studs on each side of the opening serve this purpose.

one and three-eighths-inch stops nailed to the jamb. For outside jambs, wood in which a cut has made to allow one piece of wood to rest on it, is used to hold the door. This technique is called *rabbeting.*

How to Make and Set Door Jambs

Although constructing jambs is simple, if you have some experience in constructing special joints it is helpful. Level techniques are also required. It's a good idea for the inexperienced person to practice with scrap wood before constructing door jambs.

- Carefully make the rough opening. It should be constructed two and one-half-inch larger each way than the size of the door to be hung. This allows space for the jambs, wedging and clearance for the door to swing.
- Determine any variation in floor heights at the point where the jambs rest on the floor by leveling the floor across the opening.
- Install the head jamb using *dado joints*. These are rectangular, square-bottomed grooves cut in one piece of wood to hold another piece. These joints add strength to the joining of the head jamb and side boards. Refer to Fig. 6-11 to see how the head jamb will be positioned within the dado cut. Cut the head jamb so that both ends are square. Be sure to allow for the width of the door plus the depth of the dado cuts on each side where the jamb will be positioned. Allow an additional three-sixteenth of an inch for door clearance.
- From the lower edge of the dado cut on the side wood, measure a distance equal to the height of the door plus some clearance. Then mark the wood and cut square.
- Repeat the fourth step for the other side. If the second step shows that the floor isn't quite level, be sure to cut the wood longer if the floor is lower on this side or shorter if the floor is higher. The difference in length should be exactly the same as the difference in floor height.
- Using 8-penny common nails, fasten the jambs and jamb heads together. Hammer the nails through the dado into the head jamb.
- Set the jambs into the opening.
- Floors are made with subfloors over which a finished floor is layed. So that the door will be the correct height above the finished floor, place small blocks under each jamb on

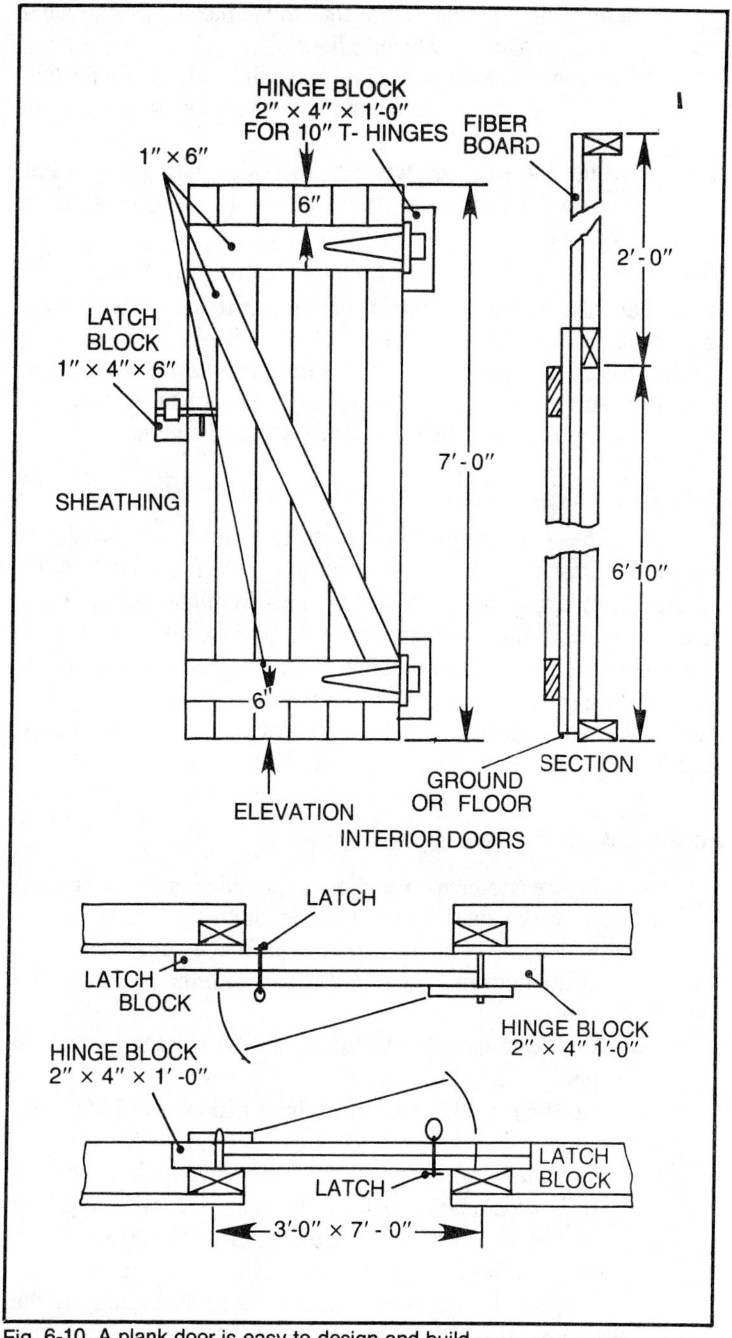

Fig. 6-10. A plank door is easy to design and build.

the subfloor, make sure that these blocks are the same thickness as the finished floor.

- Plumb the jambs and level the jamb head. This is particularly important if problems with the hung door are to be avoided.
- Align the sides by wedging the sides with shingles between the jambs and studs. Then nail the sides securely into place.

Note: For proper installation, care must be taken not to wedge the jamb unevenly. An easy way to prevent this is to use a straightedge that is 5 or 6 feet long and place it inside the jambs. Once all steps are completed it's a good idea to check the jambs and head carefully. If jambs aren't plumb, they will swing the door open or shut depending on the direction in which the jamb is out of plumb.

Door Trim

Once the door frame is constructed, treatment is needed to give it a "finished look". That's the job of trim material, called *casing*. Trim is generally available in sizes from one-half to three-quarter inches in thickness and from 2½ to 6 inches wide. If walls are made of plaster, a trim that has a concave back is used so that uneven plaster can be hidden. If the trim is installed using mitred cuts at the top corners a mitre box, mitre square, hammer, nail set and block plane are needed for proper installation.

How to Case Door Openings

- Leave one-quarter inch from the edge of the door jamb to the casing along the entire jamb.
- Select one of the side casings and cut it square and even at the bottom so that it fits flush with the bottom of the jamb.
- Allow one-quarter inch extra and cut the top or mitered end.
- Starting at the top and working downward, nail the casing onto the jamb. Be sure that the top is even with the one-quarter inch margin line. Four-penny finishing nails should be used along the jamb side, while 6-penny or 8-penny case nails should be driven along the outer edge of the casings. Be sure that nails along the outer edge are long enough to pass through the casing into the studs.

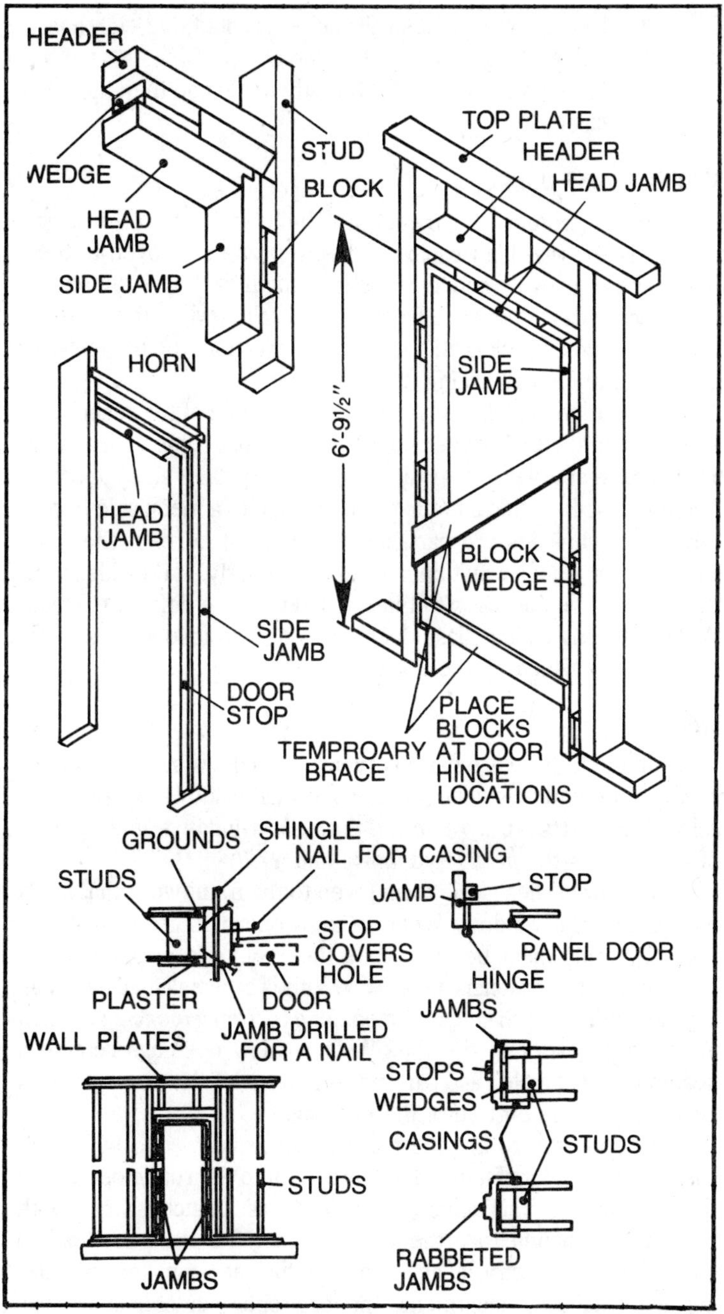

Fig. 6-11. The details of door jambs and trim.

- Use a nailset to push all nailheads about one-eighth inch below the surface of the wood.
- Position the casing for the other side and then the head casing.

SLIDING GLASS DOORS

Modern homes are frequently built with patios or large backyards. So that landscaping and views can be enjoyed to their fullest and to provide convenience in going from home to patio or yard, sliding flass doors are frequently installed when the house is built. If a patio or deck is added after construction, the homeowner will frequently install sliding doors himself.

Installation of sliding glass doors makes use of the techniques already described. However, most sliding doors come with a metal frame that is screwed into position within the proper sized opening. The doors slide within a metal track and fit into a metal side frame. Leveling techniques are particularly important here. If the track isn't level, the door won't open or close properly. If the side frame isn't plumb, the door won't fit into it and it will be difficult to close. When installing sliding glass doors, it is best to follow the specific directions provided by the manufacturer.

WINDOWS

Although there are various types of windows such as *sliding, double-hung* and *casement* (Figure 6-12) all windows consist basically of two parts—the *frame* and the *sash*. Window frames consist of four parts—the *head*, two *jambs* and a *sill*.

Window sashes is the term given to the framework that holds the glass in the window. When window openings are provided in framing, studs must be cut. Since this results in weakening the framing, reinforcement is necessary. This is accomplished by doubling the studs on each side of the opening to form *trimmers* and then inserting a header at the top. Wide window openings require a header that is doubled and trussed. A bottom header or rough sill is positioned at the bottom of the opening.

Window frames are the parts of the window into which the sashes are fitted and hung. They are set into the rough opening in the wall framing. This opening is made at least 10 inches larger both in width and height than the *window pane*, the glass used in the window. These extra inches allows sufficient room for weights, springs, balances, adjustments and for proper plumbing and squaring. For example, suppose you wish to install a window with a pane

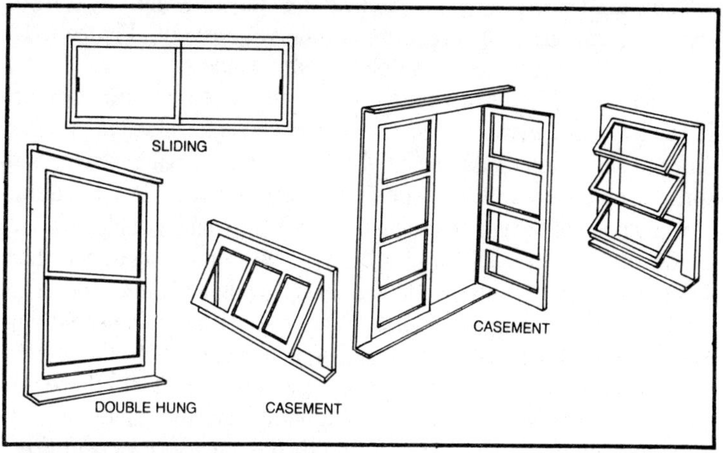

Fig. 6-12. Different types of windows.

that is 36 × 48 inches. Add 10 inches to the width and another 10 inches to the height, resulting in an allowance of 46 × 58 inches. For double hung windows, the height allowance is added to the combined pane height, not each one separately. To save on construction costs and as a convenience to homeowners who aren't craftsmen, builders today frequently use metal channel windows rather than those with weights. When this type of window channel is used, only room for the channel need to be provided—rather than the extra 10 inches.

Double-Hung Windows

Double-hung windows are particularly common on older homes. Occasionally, they require adjustment, repair or replacement. The following description of this type of window will assist you in making repairs. However, some degree of skill in the construction of different types of joints is required if a new window is to be constructed.

The double-hung window consists of two parts, upper and lower sashes, that slide past each other moving upward or downward. They are designed so that screens can be fitted on the outside to provide protection against insects in nice weather and so that storm windows can be fitted to provide additional protection against wind and cold in chilly climates. Combination storm and screen windows are frequently added in changeable climates. By closing part of the window, air conditioning units can be positioned. This type of widow is versatile and convenient. However, only half of the window area can be used for ventilation at any time and any breeze that isn't traveling toward the window will be lost. Figure 6-13 details the construction of a double-hung window.

The actual frame of a double-hung window is really a fancy box. It has a top piece called a *yoke*, two sides pieces called *pulley stiles* and the sill. Dado joints are used to position the rough casing, and form a rough box with the opening toward the studs and headers. The purpose of the rough casing is to provide a nailing surface to the studs and headers. When plaster walls were more common than they are today, the studs and headers provide a plaster stop. The outside rough casing provides a stop for sheathing. The sheathing should fit flush against it and building paper should lap the joint.

When weights are used to balance the window sash, there is a two inch space between the framing studs and the pulley stile to allow sufficient room for them. A thin strip, known as the *pendulum*, separates the weights for the two sash units. In order to provide easy access to the weights, there is an opening in the stiles near the ill. This opening contains a removeable strip and this strip is part of the stile and channel for the lower sash.

As the lower detail in Fig. 6-13 shows, dadoed into the yoke and stile surfaces is a parting strip. This strip is removeable so that the upper sash can be removed. It also forms the center guide for the upper and lower sash. The outer rough casing projects slightly beyond the stiles and yokes to form the outer guide. The inner guide for the sash is formed by a strip or stop and generally has a molding form on the inner edge. This stop is removeble so that the lower sash also can be removed.

Two pulleys are used with this type of window, one for each sash. They are mortised flush with the stile faces at the upper parts of the stiles. Either weight cord or chain can be used with them.

The sill, which slants downward and outward, is actually part of the box frame. To prevent water from being blown under the sash, one or two quarter-inch *brakes* are used. One brake is positioned at the point where the lower sash rests on the sill and the other should be placed near the outer edge to form a seat for window screens or storm sash. Near the outer edge of the underside of the sill there should be grooves into which building siding can fit. This forms a watertight seal around the window.

The room side of the sill contains an assembly called the *stool*. It contains a rabbet on its underside into which the sill fits. The stool edge projects from the wall to form a horizontal stop for the lower sash. The stool itself is part of the window trim and is made of side

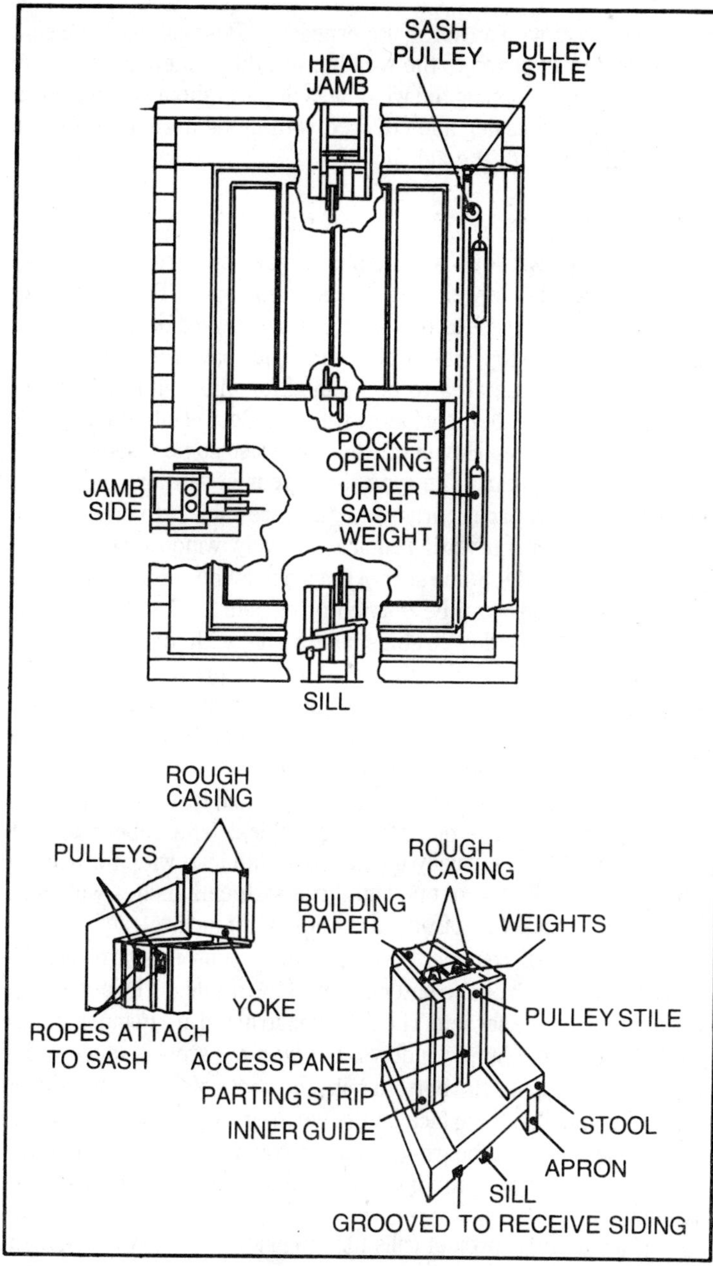

Fig. 6-13. A double-hung window which uses pulleys. Some experience in the construction of different joints is recommended before you attempt to build this type of frame.

and top casings with an apron underneath it. Positioned on the side of the window exposed to the weather are the framed finished side and top casings. Topping the window, a drip cap should be placed on the outside head casing and covered with metal flashing to form a watertight seal with the siding.

Casement Windows

Casement windows are simple to make and are considerably less complicated than weighted double-hung types. They consist of simple frames and sashes that swing either inward or outward and have hinges at the top, sides or bottom. They have the advantage of allowing their entire area to be opened for maximum ventilation. Any breeze parallel to the window will be deflected into the room by the slanting window. However, there are also some disadvantages associated with them. The type that opens inward are difficult to make watertight, particularly when they are subject to driving rain. Those that open outward require that any window screens be placed on the inside. Some sort of device must be cut into the frame so that the window can be opened or closed.

Casement window frames are made with one and three-quarter -inch planks. Rabbets are cut into the planks to receive the sash. It's a good idea to provide an additional rabbet for screens or storm sash. The first rabbet should be made one-half inch and either width depends on whether the sash is one and three-eighth inches width depends on whether the sash is one and three-eighth inches or one and the three-quarter inches thick. The rabbet for the screens or storm is usually fifteen-sixteenth inch wide for a seven eighth-inch thick sash, or one and three-sixteenth inches wide for a sash with a thickness of one and one-eighth inch.

The type of casement window that opens outward has the rabbet for the sash on the outer edges of the frames. The inner edge is rabbeted for the screen. The sill is constructed in a manner similar to that for double-hung windows, but with the stool is much wider and forms a stop for the bottom rail. The width of the frames should extend to the sheathing face on the weather side of the window and to the plaster or wallboard face on the room side. Fig. 6-14 shows the assembly of both the inward and outward openings casement windows.

If a particular project calls for the placement of two casement window in a row on one frame, they can be separated by a vertical double jamb called a *mullion*. If you prefer, you can place two stiles together in somewhat the same manner as a french door. The edges

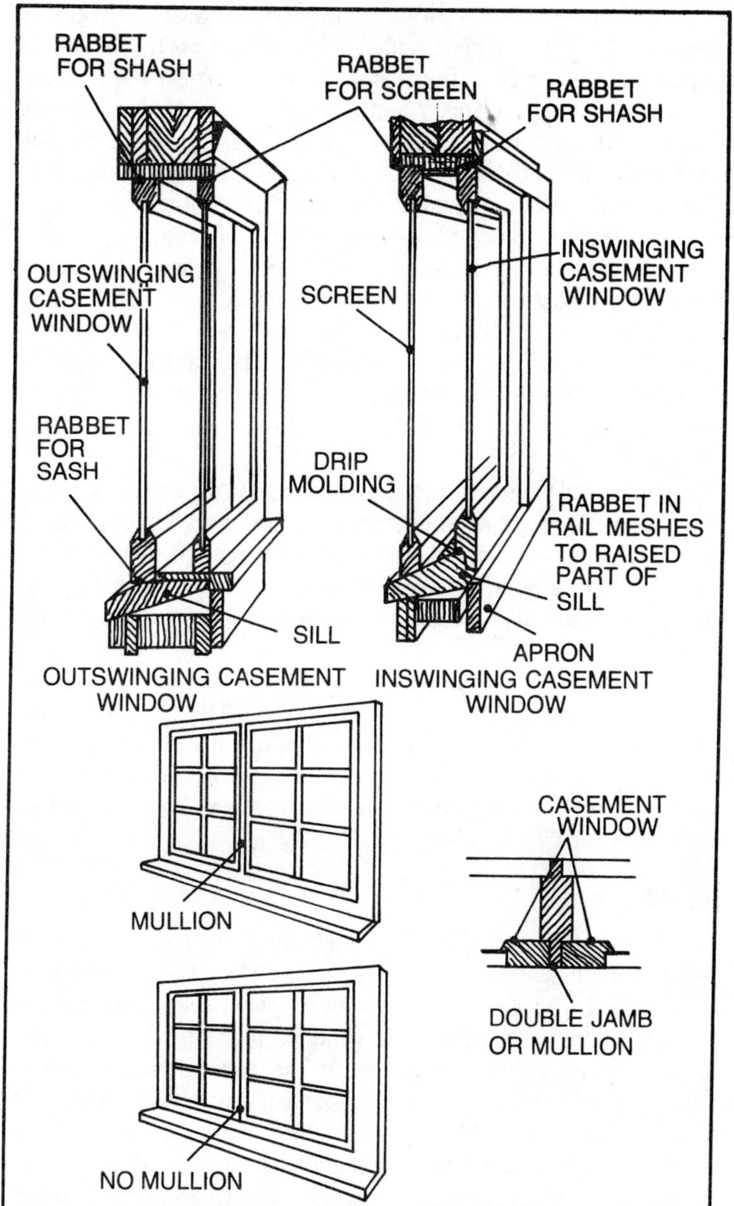

Fig. 6-14. The construction of inward and outward opening casement windows. When two such windows are constructed in the same frame they can be separated by a mullion or the stiles can be placed together like a french door.

of the rabbet can be formed in different ways. They can be a reverse rabbet, a beveled reverse rabbet with battens such that one is attached to each stile or they can be beveled *astragals*—the term given to T-shaped molding. When astragals are used, one should be attached to each stile. Battens and astragals are used to give improved resistance to weather. Astragals are more resistant to loosening as the windows are used. Figure 6-14 shows two pairs of casement sashes in one frame hinged to a mullion in the center.

The same illustration shows that inswinging casement window frames are similar to the outswinging kind with the sash rabbet cut in the inner edge of the frame. Notice that the sill contruction is slightly different. It's made of one piece with a rabbet toward the front edge for a screen or storm sash. The back is raised where the sash rail sits. In order to form a stop for the rail this surface is rabbeted.

As with all windows, moisture and exposure to weather poses a problem. To prevent water from entering, the sill of a casement window should slope downward 1 inch for each 5 inches of width. The sill also forms a base for the outside finished casing and is designed to extend about one and one half-inches beyond the sheathing.

There is another difference between the inward swinging and outward moving types of casement windows. The bottom sash rail of the inswinging type has a bottom edge that is rabbeted to mesh with the rabbet on the sill and a drip molding set in the weather face. The purpose of this drip molding is to prevent rain from being blown under the sash.

FLOORS

To anyone contemplating the addition of a room or the modernization of a house, constructing a floor can be a difficult problem. However, if you have gained some experience in construction techniques this project is most routine. If the project is a new addition or an existing structure or the construction of a new building, the floor is made after the foundation and basic framework are completed. For existing rooms, the project becomes more complicated since it is first necessary to remove existing flooring when a major restrucuring is being done. Such a project is best left to experienced individuals.

Once the basic framework has been completed, the subfloor is laid diagonally on the joists and nailed into position with 8 or 10-penny nails. The floor joists forms a framework for the subfloor.

The boards that fit upon them should be at least 8 inches wide and should be fastened to the joists using 3 or more nails per joist. Of course, the subfloor can be over 1 inch thick. In such cases, larger nails should be used. Figure 6-15 shows the relationship of the subfloor to the joists and how the finish floor fits upon it. When a house is being built, the subfloor is poisitioned and fastened before the walls are framed. The subfloor can then be used as a platform upon which the carpenter can work.

Once the subfloor is in position and the outside of the addition has been completed, it is time to lay the finish floor. This might be hardwood or a host of other types. Usually square-edged or tongued and grooved types are used when the finish floor is to be wood, but fancier varieties can also be layed. Figure 6-16 shows methods for nailing square-edged flooring. Note that when positioning the finish wooden floor on a subfloor, building paper is used to keep out dampness and insects. If the addition is to be part of a storage area, remember that the wood floors must be strong enough to carry the load. This is what determines the thickness of the wood to be used, the sheathing and the approximate spacing of the joists. Figure 6-17 shows the methods for nailing tongued and grooved flooring.

Although a carpenter works primarily with wood, sometimes it's useful to know how to work with other materials as well. As a

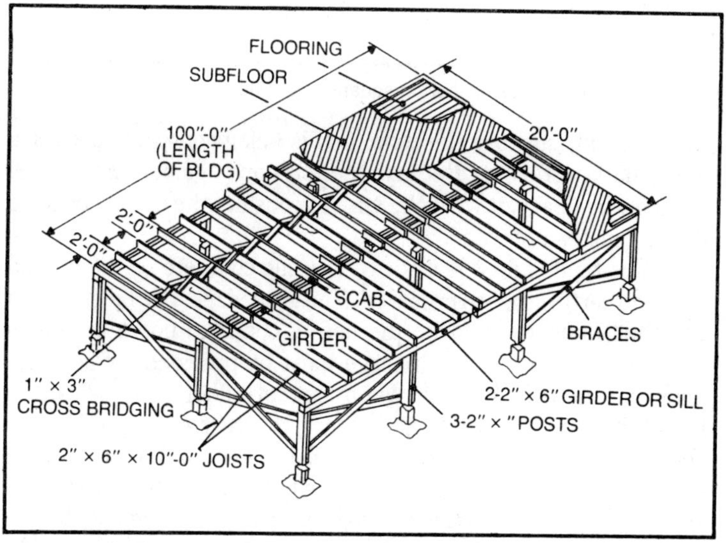

Fig. 6-15. When a floor is placed on a foundation it would be installed like this. Note the position of the floor on the subfloor. The dimensions shown are typical for braced piers, sills, girders and joist construction.

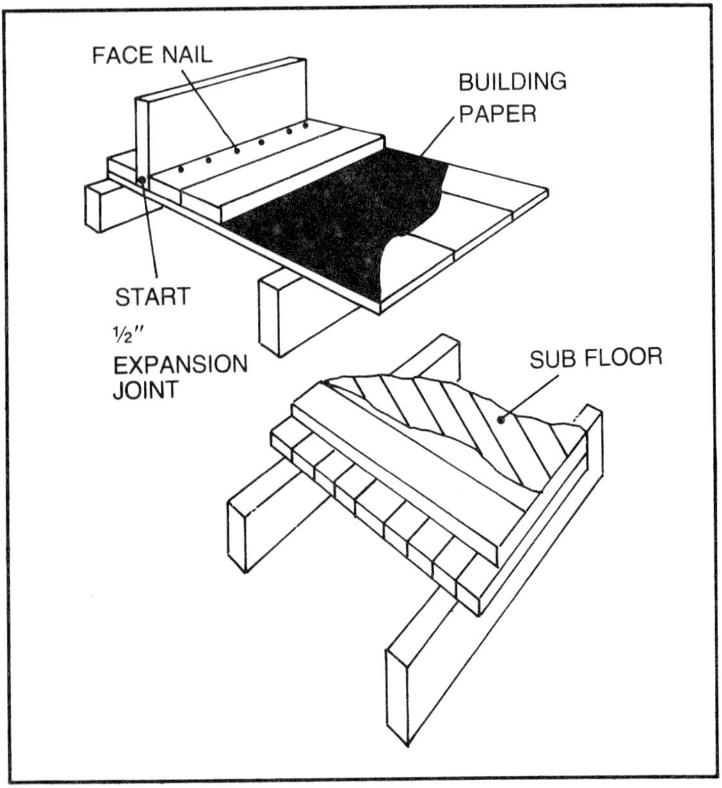

Fig. 6-16. The methods for nailing square-edged flooring.

case in point, suppose the room that is being constructed is for a heavy duty workshop. Because the tools and equipment used will probably be heavy, it would be wise to consider the use of concrete flooring instead of wood.

This type of floor is made by pouring concrete on the ground after the earth has been prepared. The ground must be been graded and then tamped. Because the use of concrete floors promotes dampness, some sort of drainage must be provided for both the floor and the area near it. When concrete floors are planned, a foundation wall is poured first. Once the rest of the building is completed the floor can be poured. This allows you to protect the concrete floor while it sets. If you plan to use this type of construction be sure to check on local ordinances and requirements before starting the project. Sometimes specific techniques must be used.

When you pour cement a special form must be constructed to act as a type of container. As the concrete dries, it conforms to the

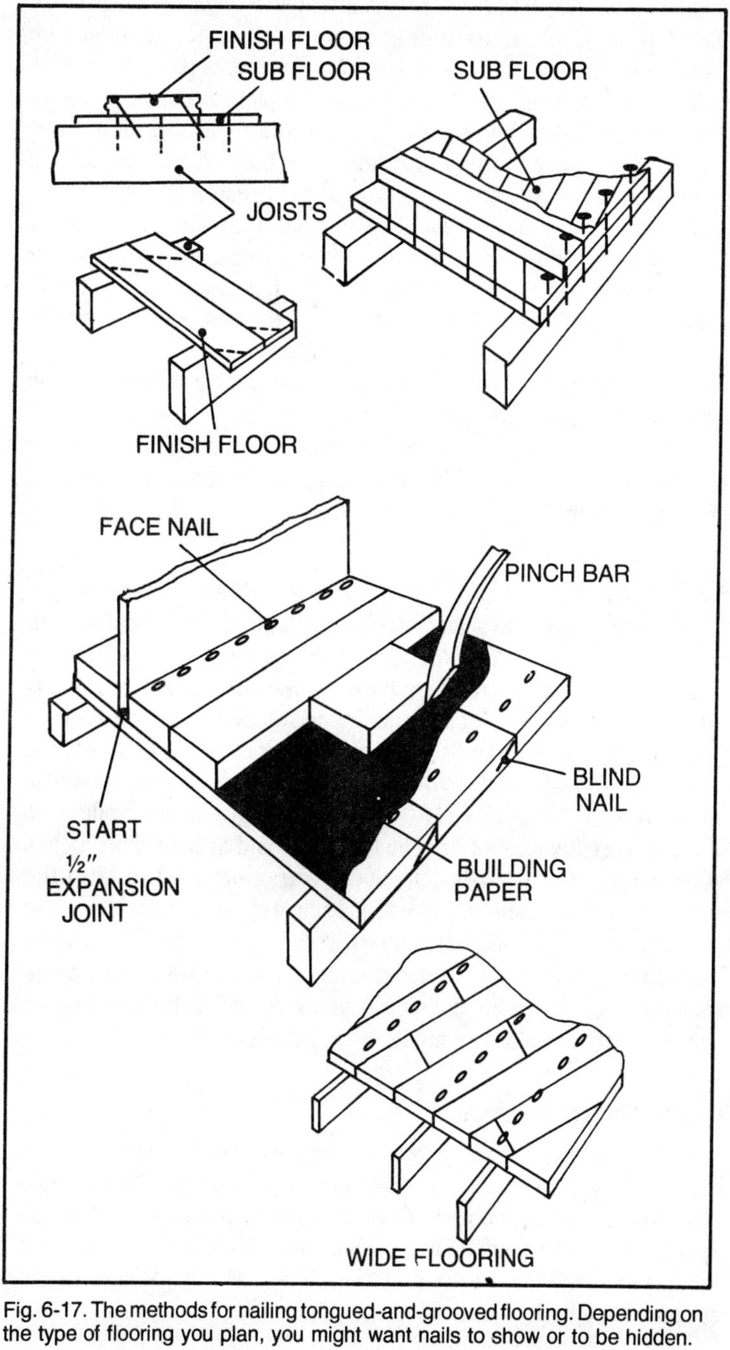

Fig. 6-17. The methods for nailing tongued-and-grooved flooring. Depending on the type of flooring you plan, you might want nails to show or to be hidden.

shape of the form. If the concrete is intended to support a great deal of weight or if extra strength is desired, the concrete should be reinforced with special rods or conduit. The forms protect them as well as the concrete until hardening is completed. Since the forms must be tight, rigid and strong to prevent against loss of mortar, loss of water or streaking and honeycombing, the placement of forms is particularly important. For the best results it's wise to get assistance from someone with experience in this area.

Although you might think of a wood flooring before any other type of floor, there is a host of floor types that can be installed to fit specific circumstances. For example, an earthen floor can be the cheapest and easiest floor to install in an outdoor shed. Brick can be used in a patio or pool area. Investigate plans and discuss them with building material suppliers. A supplier can advise you on the best techniques to use. But remember, much of the enjoyment from carpentry comes from the opportunity to be creative. Don't be afraid to experiment.

STAIRWAYS AND STEPS

Building stairways and steps might seem complicated, but constructing them isn't as difficult as it appears. Stairways have framing on the sides, called *stringers* or *carriages*. It is this framing that determines how steep the stairway will be. While the average person refers to the planks that are used for climbing as steps, the carpenter knows them as *treads* that sometimes have additional pieces toward the back called *risers*. The stringers are made from material that is either 2 or 3 inches thick and at least four inches wide. As Figure 6-18 shows, blocks are often used to form the steps. Three stringers are used for each step, one at each end and the third in the middle. When a new stairway is installed, the floor joists must be properly framed around the stair well so that there will be enough room for the stair framing as well as for the finished trim that will be installed along the staircase.

How to Construct Treads

Treads are the wood that is stepped upon when stairs are ascended or descended. They might consist of little more than rough planks or they can be quite elaborate. In many areas, stairway specifications are detailed in local building codes. Be sure to check these requirements before strarting. The following type of tread support is typical of stair construction:

— Stringers are made of 2 by 4 inch lumber with triangular blocks, constructed from 2 by 6 wood and nailed to one edge. The amount of wood depends upon the number of steps and the height of the stairs. The first step is to plan these factors and obtain sufficient wood to construct the stringers.

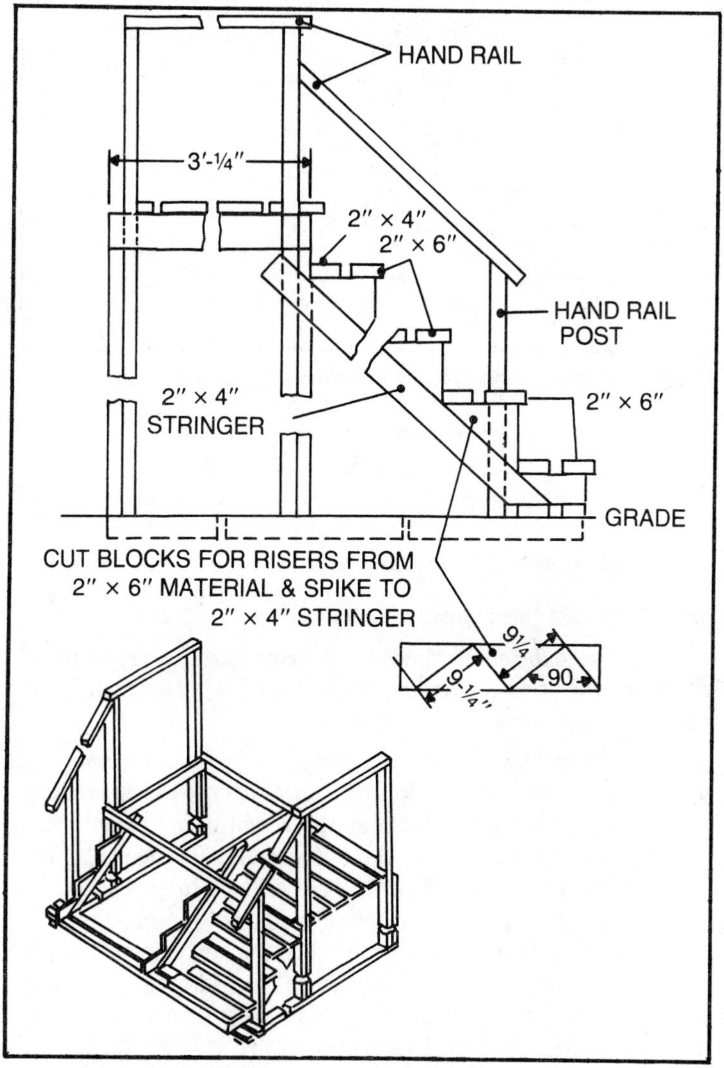

Fig. 6-18. The details of step construction. Actual design and finishing materials can vary depending on the purpose and location of the stairway.

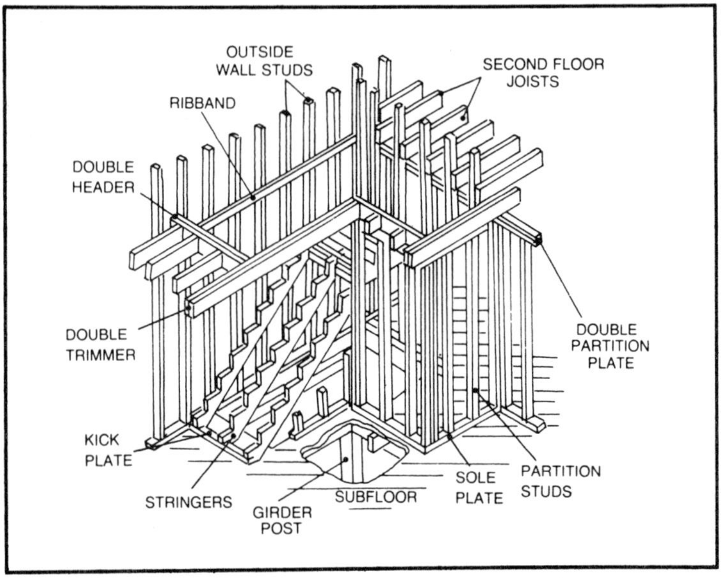

Fig. 6-19. A stairway showing building frame and stringers.

— Cut the blocks from the 2 by 6s.
— Nail the blocks to the stringers (Fig. 6-18).
— Fasten the step stringers at the top and bottom as shown in Fig. 6-18B. Once this is accomplished, the stairway frame should be completed.

How to Complete the Stairway Frame

There are different methods for framing stairs. The type to use depends on the style of stair installed. The following example is for simple, straight stairs.

— Obtain a piece of simple, straight narrow stock that is as long as the distance from the lower floor to the upper floor level. This is the same as the height of the lower room plus the thickness of the floor joists, plus the rough and finished flooring. This stock is called a *story pole.*
— The average distance between stairs is generally about 7 inches. Mark 7 inch intervals on the story pole. Note: If the story pole can't be marked in 7 inch intervals, try using spacing that is slightly larger or smaller than this distance. Keep as close to 7 inches as possible. Several adjustments might be necessary before even spacing is obtained. The number of spaces will be the number of risers in the stairs.

— Look at the triangle drawn in Fig. 6-20. The figure can be thought of as representing an entire flight of stairs. The vertical side is called the *rise* of the stairs. The horizontal side is referred to as the *run*. Measure the length of the wellhole for the length of the run of the stairs. If you are contructing the stairs from pre-printed plans they will also show this distance.

— Obtain the width of each tread. This is done by counting the number of risers and subtracting one, because there is always one more riser than tread. Divide this figure into the run of the stairs. The result is the thread width.

— Refer to Fig. 6-21 and Fig. 6-22. Use a steel square to lay off the run and rise of each tread and riser on the stringer stock.

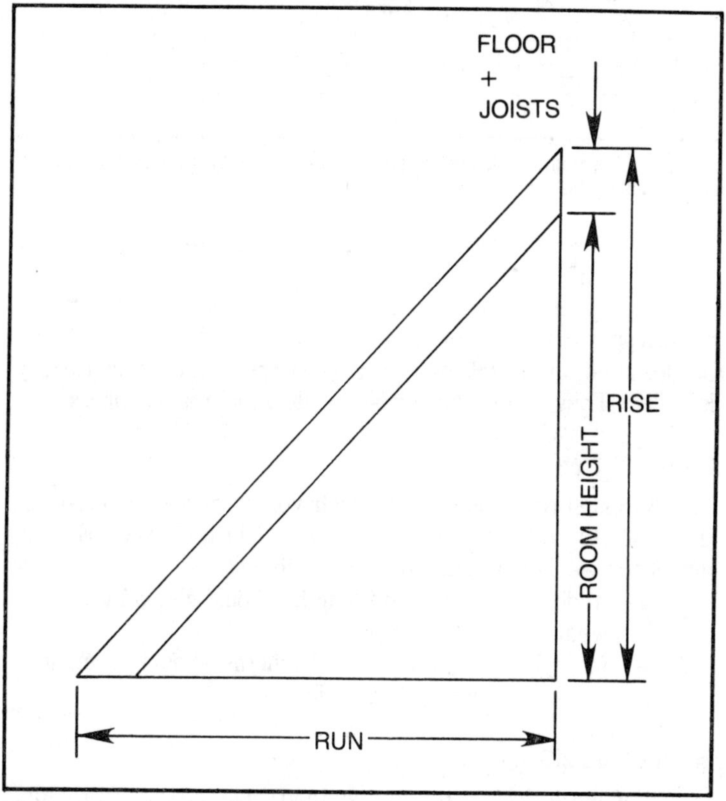

Fig. 6-20. The stair triangle shows the relationship of the rise of a staircase to the run. Note that the rise is the height of the lower room plus the joists and floor of the upper room.

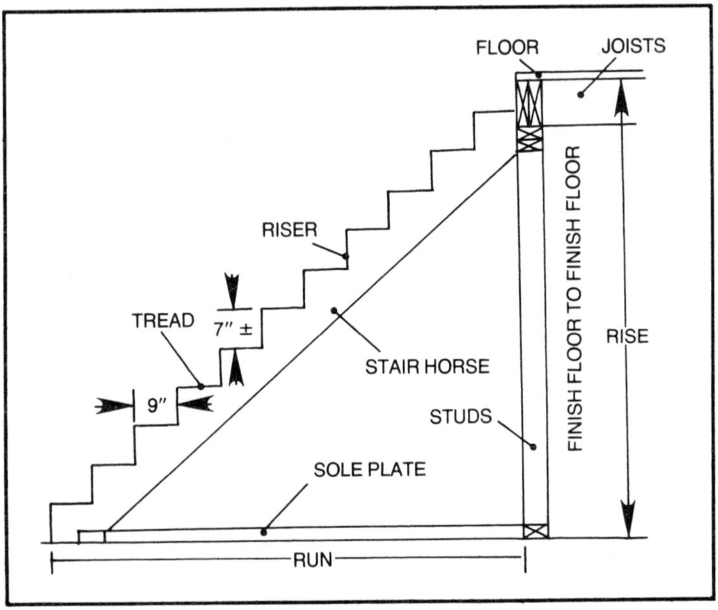

Fig. 6-21. The method of laying the stair stringers. Note the relation of the risers and treads to the stair triangle.

— Assemble the steps as shown in Fig. 6-18 through Fig. 6-22.

The finishing touch on stairs is the handrail. A handrail can be made from two by fours or elaborately carved and polished rails can be designed and installed. Iron railings are attractive and easily installed. They are available from building material suppliers.

Riser and Tread Tips

Although risers and treads can have a variety of dimensions, the ideal step run and rise totals about 17 inches. The following guidelines will help in planning comfortable steps.

— Adding riser and tread lengths should give a figure between 17 and 19 inches.

— Multiplying riser and tread lengths should result in a number between 70 and 75 inches.

WOODEN DECKS

Wooden decks are attractive and add value and style to a home. Because they are generally above the ground, air can circulate freely so they tend to be cooler than other patio areas. They add

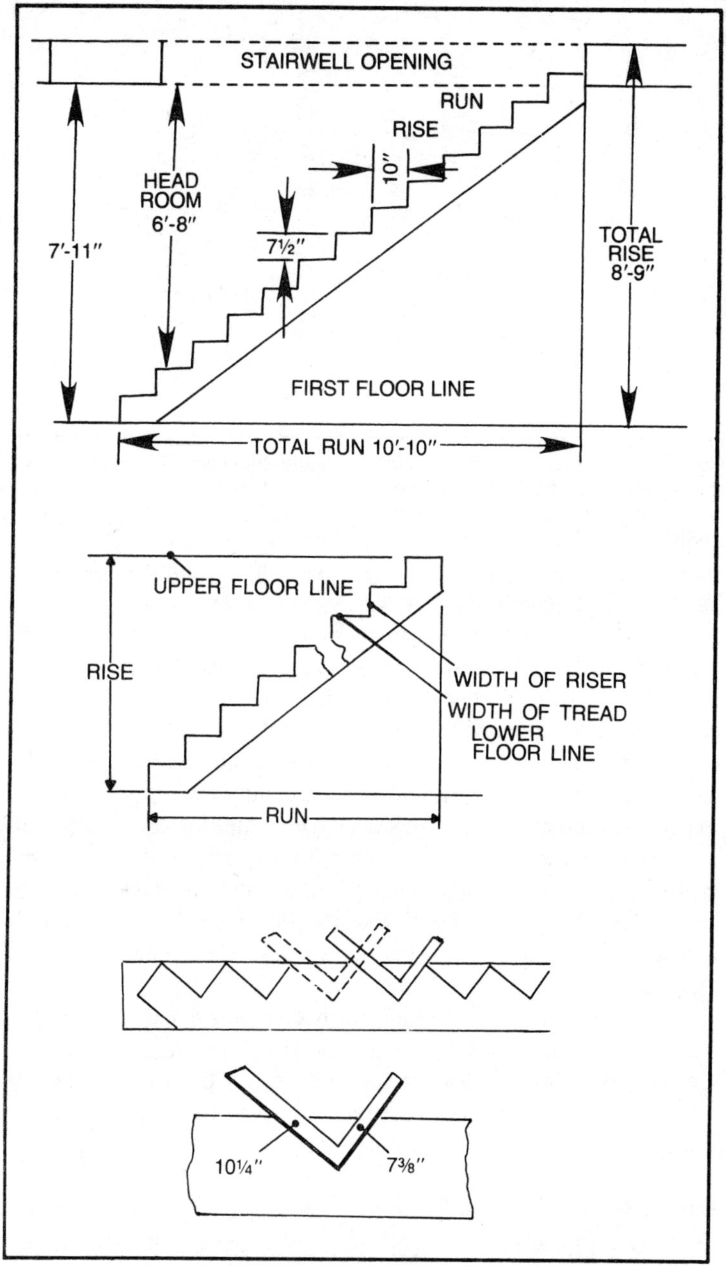

Fig. 6-22. The components for stair construction. Although actual distances might vary, a typical rise of 8 feet 9 inches and run of 10 feet 10 inches are shown.

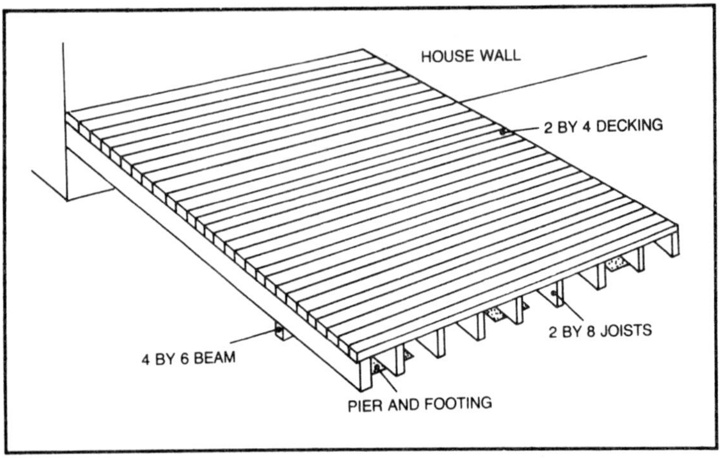

Fig. 6-23. Decks can be designed in many different ways. The simple deck shown here could be used on a level area.

additional usable space to a home and can serve as an outdoor living room. For these reasons, many beginning carpenters try their hands at building a deck. However, a special caution is warranted. Although the techniques used in deck construction are simple to the individual with some experience, there are additional factors that must be considered. Local codes might prohibit the addition of decks. Where they are permitted there could be maximum size construction techniques and materiais specified. Permits are usually required before construction can be begun. Once you've determined local requirements, deck design becomes important. Special skills and techniques, such as site clearing, preparation and concrete use, are frequently required. But these obstacles can be overcome. Permits can be obtained and land can often be cleared with a shovel, wheelbarrow and hard work. The results are often very rewarding.

The use of concrete footings provides a sturdy support for the deck. The easiest way to do this is to use premixed concrete. Create footing holes in accordance with local codes. Mix the cement with water so that it forms a paste and then pour the mixture into the footing holes.

How to Make a Wooden Deck

- Figures 6-23, 6-24, and 6-25 detail the following steps. Although the land over which the deck will extend doesn't need to be level, flat land makes construction

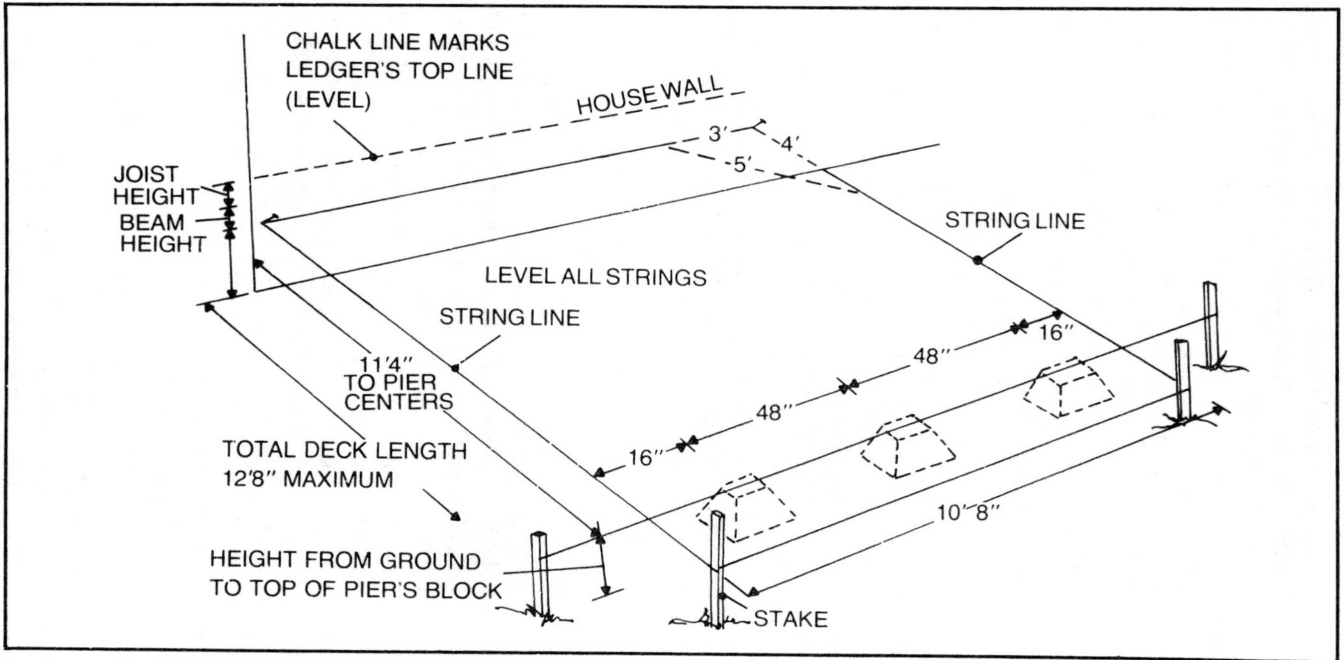

Fig. 6-24. String is used for determining exact location of the deck and foundation and for leveling. The use of a 3', 4', 5' right triangle assists in positioning the deck properly but it is not absolutely necessary.

easier. Either level the land or cut 4 by 4 inch posts of different lengths so that lumber joining the tops of the posts will be level.

- Determine the location of the deck on the house using string to mark boundries.
- Decide on the length and width of the deck.
- Mark the width and height of the deck on the house wall.
- If you will have a door opening onto the deck, make sure that the height of the deck will be even with the floor of the room inside the house.
- From one end of the width mark on the wall, measure the thickness and mark this distance on the wall.
- Repeat this step at the other end of the width mark.
- Drive two temporary nails part way into the house wall, one at each mark made in the previous steps.
- Tie a string between the two nails.
- Use a level to make sure the string is level. If necessary, make adjustments. Double check to make sure the string is level. If necessary, make adjustments. Double check to made sure that the deck height won't be higher than the interior floor if there will be a door opening onto the deck.
- Snap a chalk line along the length of the string. The top edge of the supports will align with this line.
- The next step is to layout the height of the supporting lumber. From each temporary nail, measure down the height of the joists (7¼ inches) plus the height of the beams (5½ inches) and again make marks. Drive temporary nails partially into them.
- From each nail, project down to the bottom of the wall and make another set of marks.
- From the marks at the bottom of the wall, measure out across the land until the desired length of the deck is reached. Then drive a stake about 3 feet into the ground at each spot. This marks the outer corners of the deck.
- Check to make sure that the stakes are positioned properly. If you were to draw a line from each stake to the house the lines should meet it at right angles. Otherwise, the deck won't meet the house properly.
- Square the stakes.
- Tie a string tightly from each nail to its corresponding stake and then level each string.

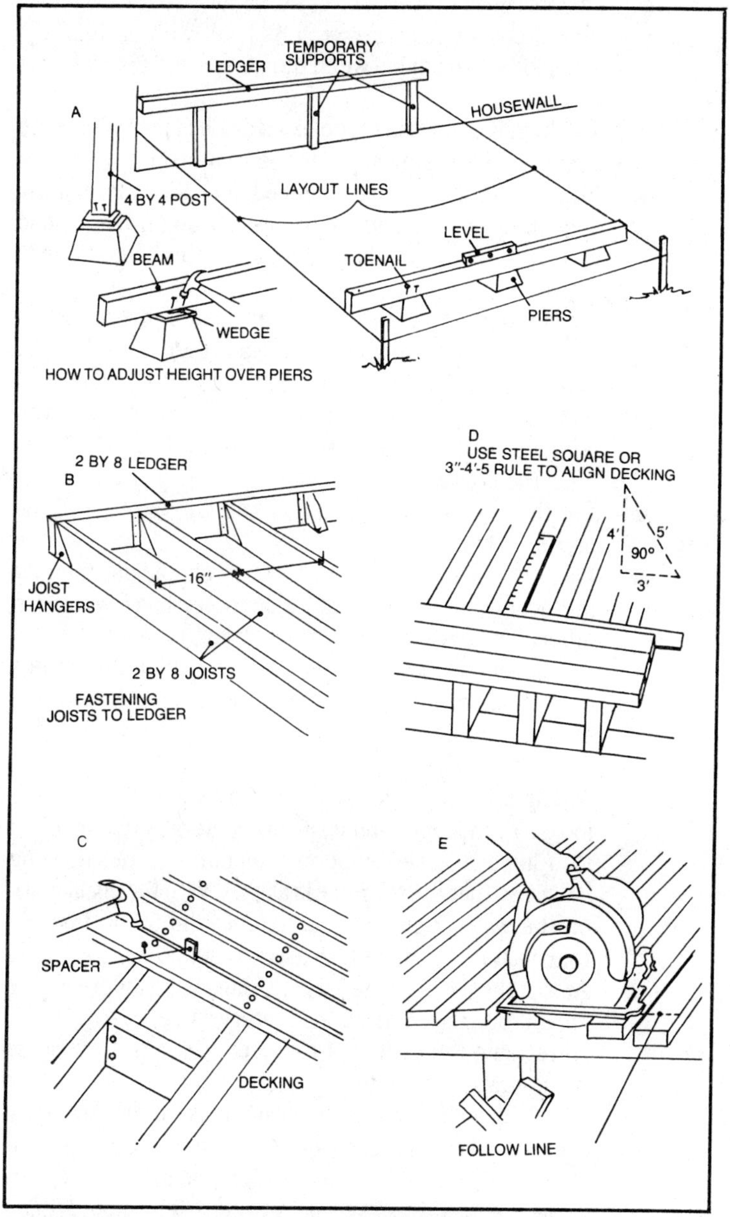

Fig. 6-25. How the deck is pieced together and fastened to the ledger and house wall. Fasten the ledger to the wall and the beam to the piers as in A. B shows how the joists are attached to the ledger using joist hangers. Cross bracing and spacing is shown in C and deck alignment using a 3′, 4′, 5′ triangle or square is shown in D. Edge trimming and sealing is shown in E.

- Double check the leveling of the stakes. If the leveling is accurate the two strings will be level with each other. Tie another string between the stakes and check that it is level.
- The height of the top of the block on each pier where the beam will rest is marked by the two strings.
- Although there is generally enough space between the decking boards for water to pass through, it's a good idea to plan a slight slope away from the house to help runoff.
- Since the piers support beams and posts, they must be sturdy. Determine where you want them and then dig holes for footings. Be sure that these holes are deep enough to supply the needed support when filled with concrete and that they satisfy any local codes. Then pour the concrete.
- Let the concrete dry until it is stiff enough to hold the piers.
- Place precast piers on the footings and adjust them to the desired height so that all piers are level with each other. To check heights, run a string between the two stakes so that the string is barely touching the strings extending from the wall. The string should pass just above each pier.
- Allow the concrete to dry thoroughly.
- A ledger is used against the house wall to support the joists. Both ledger and joists are generally two by eight inch lumber. Use temporary supports to position the ledger so that the top edge is aligned with the chalk mark on the wall. Then attach a beam to the piers. Four by four inch posts should be used to support high decks.
- Attach the joists to the ledger using joist hangers. Joists should be spaced at 16 inch intervals.
- Determine the center of the deck.Using two by eights cross-brace the joists.
- Nail decking material to the joists, making sure that they are aligned at right angles to the joists. A good trick to insure uniformity between decking boards is to determine the desired spacing between them and fashion spacers of this thickness. After the first board is fastened to the joists, lay the next board on the joists against the spacer and then nail it into position. Repeat this for each deck board.

- Trim the edges of the deck boards if necessary so that they are all uniform
- Seal the ends so that they are protected from the elements.

Once these steps are completed, you are ready to add any railings or stairs and then stain or paint the wood to enhance the beauty of the job and protect the wood.

CEILINGS

Once a room has been built or decorated, its ceiling provides the final touch. Whether simple hardboard or ornate-carved wood, the ceiling in a room accents lighting, the wall covering and finished floors. In spite of the potential advantages of a fine ceiling, they are often ignored, in need or repair or replacement or just plain ignored. The time spent installing a new ceiling or repairing an old one increases the value of a home. With modern materials, almost anyone can install a new ceiling.

There are many types of ceilings that you can easily install. *Suspended ceilings* are panels made from any number of materials and in a wide variety of designs. They are positioned within frames that are hung from supports. *Ceiling tiles* are light, decorative and easy to work with. They are nailed or stapled to furring strips placed across joists or an existing ceiling. *Wallboard ceilings* are applied in basically the same manner as walls and can be painted. *Wood ceilings* can be created from plywood, boards, or paneling.

Ceiling Frames

Perhaps the most important part of any ceiling, is the portion that you don't normally see—the portion above the covering. Called *ceiling frames*, the material and construction of the frame depends on the location of the ceiling in the home. Frames located in the basement differ from those on the first or top stories.

The most important part of the ceiling frames are the joists. They must be strong enough to support the ceiling as well as any flooring in the level above. Most frequently, joists are spaced 16 inches apart and run across the narrow dimension of a building. When joist lumber is joined together to make long spans they are supported at this juncture by the top plates of a bearing wall. When furring strips are used to install a new ceiling they are positioned across the joists.

SUSPENDED CEILINGS

In office buildings, suspended ceilings are ideal because they are easy to install and they have replaceable components. Because they are hung from metal grids they provide extra room above the ceiling panels so that cables, ducts or plumbing can easily be routed to meet office space or machine requirements. In the home, suspended ceilings are a natural choice for completing a room that has an uneven ceiling, or where you want to hide joists, beams, rafters, fixtures or unsightly older ceilings.

While some suspended ceiling panels are available in 4 by 16 foot dimensions, the most commonly used size is 2 by 4 feet.

Determining the Number of Suspended Ceiling Panels

- Take a piece of graph paper or draw small squares on a plain piece of paper.
- Measure the wall lengths at the height where you want the ceiling panels to be positioned.
- Lay off the ceiling area on the graph or boxed paper. Allow one box for each square foot of ceiling.
- Lightly shade the included areas.
- Using a colored pencil, mark off blocks corresponding to the size panel you will use. Again allow one square for each square foot of paneling.
- Count the number of full areas enclosed within the colored pencil blocks. This is the number of full panels you will need.
- Count the number of squares outside the colored regions. This is ceiling area for which you will have to cut additional panels.
- Use a different colored pencil and trace boxes so that the opposite borders of the room are equally trimmed panels. This will give a more professional looking result when the ceiling is installed.
- Usually, panels installed around the perimeter of the ceiling have to be cut to fit. Determine the size of these panels by measuring the space from the last full row of panels and take one half of this figure. This gives the size of the panel to be placed against one wall and also the opposite wall. Repeat this calculation for other room dimensions.

How to Install a Suspended Ceiling

Once the required amount of material has been determined, the next step is to decide if you want to have illuminated panels. These are similar to the decorator panels except that they are constructed to allow light to pass through them when they are placed over areas where there are lighting fixtures. Shade the areas on your graph paper where fixtures are positioned. You are now ready to install the ceiling.

- Calculate the height of the ceiling. The ceiling should be placed at least 5 inches below any lights and 3 inches beneath plumbing.
- Make a chalkline around the room at the height determined in the first step.
- Install L-shaped angle molding so that its base is at the chalkline.
- Install the main runners. For new ceilings, they are positioned perpendicular to ceiling joists. For covering existing ceilings, position them along the longer dimension of the house. Determine the required length of these runners and cut them to this figure with tinsnips. Then rest them on the molding at each end.
- Support the runners every 4 feet with 12 gauge wire. Where the joists are accessible, attach the wire to them. Where joists are not visible, first determine their location by carefully tapping nailholes through the existing ceiling until solid contact is made. Be sure to avoid wiring, plumbing or weakening the existing ceilings. Once one joist is located, other joists are normally about 16 inches apart. Screw heavy-duty screw eyes into the joists and attach the wire to them.
- Position 4 foot cross tees. Lock them to the main runners by bending their tabs after they have been inserted into runner slots.
- Push the full panels into place.
- Cut partial panels to size with a utility knife or fine saw.
- Push partial panels into place.

CEILING TILES

Ceiling tiles provide an easy way to renew a ceiling or provide additional soundproofing in play areas. They are available in rectangles but they are most frequently sold in 1 foot squares. This type of tile comes in a wide assortment of colors and patterns and can be applied directly to existing ceilings that are level and in fairly good shape. They are lightweight so they don't pull a great strain on existing ceilings. They are easy to hold in position over your head while you are inspecting them. They are stapled into position or fastened with a special adhesive. When existing ceilings are weak or not level, ceiling tiles can be placed over furring strips that are installed over the existing ceiling.

Determining the Required Number of Ceiling Tiles

The same technique for determining the number of suspended ceiling panels can be used for calculating the required number of ceiling tiles. Calculations are somewhat easier with ceiling tiles since they usually come in 1 foot square tiles. Plan on using a few extra tiles since they tear easily. Until you gain some experience in tile application you might damage a few. Since they are so simple to install and easy to remove before the adhesve sets, errors can be readily corrected.

Installing Tiles on a Flat, Level Ceiling

- Make sure the ceiling is fairly level and flat. If it isn't, use the second method.
- Begin in the corner of the room away from the doorway—this corner is the corner you notice first when you enter a room.
- Dab some adhesive on the back of one of the tiles near each corner and at its center. Don't put the adhesive too near the edge or it could spread outside the tile when the tile is applied.
- Press the tile into position at the corner of the room.
- Repeat the third step with another tile and then position it next to the first one. Ceiling tiles generally allow for an overlapping fit. Slide the second tile against the first before pressing it in position. Repeat this process until all full tiles are positioned.
- Use a utility knife with straight edge blade areas and then install them.

For new ceilings or where existing ceilings aren't level or are weak, use the following method. Note: if a ceiling is very weak it could be dangerous. You should remove it and install a new one. This method makes use of 1 by 3 inch furring strips that are applied at right angles to the joists. For covering existing ceilings, assume that the joists run along the narrow dimension of the house.

— Measure the dimension of the room at right angles to the joists and cut furring strips to this size.

— Nail the first furring strip flush against the wall at right angles to the ceiling joists.
— Use a spare tile as a ruler and position the other strips parallel to the first one so they are one tile length apart. Double check spacing by using a ruler about every 4 feet.
— Make certain that the strips are level. If they aren't, they can be adjusted by driving shingles between the strips and joists (or between strips and existing ceiling).
— Fasten the tiles to the strips by stapling the tile flanges to the furring strips. Begin with border tiles.
— Make sure that the flanges of one tile fit securely against the next and install all full tiles.
— Cut smaller tiles to size by using a utility knife.

WALLBOARD CEILINGS

Wallboard ceilings provide a solid look that can be painted to match or contrast with the walls. Wallpaper can be extended from the walls to cover the ceiling to give an unusual decorator touch. When installed, wallboard ceilings are strong and durable. However, wallboard panels are large, awkward and heavy. They must be held overhead while you are positioning them and it is sometimes difficult to place them properly while you are trying to nail them in place. Once they are installed, they have seams and nails that must be concealed to yield a smooth, even look.

Wallboard can be installed directly over joists. When it is used to cover an existing ceiling it can be applied to furring strips.

Installing Wallboard Ceilings

- Determine if the ceiling joists are level. If an existing ceiling is to be covered, make sure that it is flat and level. If joists or existing ceilings are flat and level, the wallboard can be applied directly over them. Otherwise install 1 by 3 inch furring strips 24 inches apart. Except for spacing, the procedure for this is the same as the first five steps for installing ceiling tiles. Because of the weight of the wallboards, furring strips must be firmly anchored. Whether they are positioned directly over joists or covering an existing ceiling, make sure that they are secured to ceiling joists with strong hardware.
- Because of the weight of the panels, it generally takes two people to position and fasten the panels. To make the job easier, both individuals should be at the same

height. Construct a scaffold by setting up two sawhorses with a few short planks across them. For safety, be sure that the sawhorses are sturdy and that the arrangement can support the weight of the individuals and wallboard and that it won't move as the people are working.

- Position the first panel in a corner of the room. Installers should stand at opposite ends of a panel with the panel held above them. Starting at the center of the panel, drive the nails in a pattern toward the corners. Nail outward from the center.
- Repeat this process until all full panels are installed.
- Cut remaining panels to size and install them in the same way.
- Place spackling tape over the joints and spackle compound outward from the tape to prevent high spots. The installation is complete and the ceiling is ready for finishing.

Wallboard Ceiling Tips

Many communities have codes governing the location of nails when wallboard is installed. It's a good idea to check these codes before beginning work.

When driving nails, be sure that they are driven below the surface of the wallboard. This way they can be hidden by placing spackling compound over them. Feathering the compound outward away from nails results in a professional-looking job.

If a new wallboard wall is installed, construct the ceiling before the walls. Taper the edges of the panels against the wall where they will meet the wall. Do the same thing for the edges of the wall panels where they meet the ceiling. Then apply spackling compound to the joint and feather outward from it. If you are not happy with the results, apply ceiling molding. This not only provides an extra finishing touch normally left off new construction but, it also hides mistakes where the wall and ceiling join.

SPECIAL CEILINGS

Don't feel restricted to common ceiling materials. The same techniques used to install the most popular types of ceilings can be applied to the construction of unusual or creative ceilings. For example, you can install plywood or plywood paneling or even wood planks. Some types of siding form an unusual and attractive topping for a room. Think of a ceiling as a wall on its side and almost any type of wall covering can be applied to a ceiling. Be sure that the weight of any ceiling is supported by strong hardware attached to building members.

PROTECTING THE EXTERIOR OF A HOME

Siding provides beauty and protection but might not be rigid or strong enough when nailed directly to frame members. Many types of siding don't offer much in the way of insulation. *Sheathing* is special material placed over the frame of a new house to overcome these difficulties.

There are many different sheathing materials. Plywood has long been a favorite of builders because its large size permits rapid installation and so reduces construction costs. Its lateral strength eliminates the need for diagonal bracing in wall framing.

Exterior fiberboard is increasing in popularity because it's easy to work with, is available in large panels for quick installation and offers better insulating properties than plywood and other types of sheathing. With the ever increasing awareness of the need for energy conservation and its increasingly higher costs, any material with good insulating properties should be considered for use. However, exterior fiberboard offers only fair rigidity and should be applied only with framing that has been suitably braced.

Exterior gypsumboard offers good rigidity at low cost. However, its insulating abilities are low. Diagonal bracing should be installed. This type of sheathing isn't intended as a nailing base for siding. Be sure that siding used over exterior gypsumboard is attached to framing lumber.

Special solid boards can also be used for sheathing. They offer good rigidity, fair insulating properties and can be installed either horizontally or diagonally. When it is applied horizontally diagonal bracing is required. There is a form of solid board sheathing that is available in 2 by 8 foot panels of edge-glued lumber. This type has a convenient layer of building paper.

SIDING

Siding is the material applied to the outside walls of a house and it is one of the first things most people will notice when looking at a home. Fine siding adds beauty to a home, protects it from the elements and plays an important role in the insulating process. It keeps warm air inside the house and cold air out in winter. The reverse is true when air conditioners are used in summer. Siding that is in bad repair will allow moisture to enter the house. The result could be water damage, expensive repairs and discomfort to those inside. Installing new siding adds value to a home, improves its appearance and saves money on fuel bills. Depending on the type selected, it is usually easy to install.

Facts About Siding

There are several different types from which to choose. It comes in metal such as aluminum or in different styles of wood and asbetos. Solid board siding looks attractive and it can be painted. Plywood types are strong and faster to install. Shingles and shakes are rugged and give a home a country look. Shakes are available in vinyl and asbestos materials that look like wood, but that can go for many years without requiring paint or replacement. The techniques to install non-wood types vary, so it is best to consult manufacturer's directions before you begin working. A building permit is usually required before you install siding on a house. Be sure to check local regulations.

Whether you are applying siding to a new home or you are covering or replacing existing material, you need a strong base to drive nails into. It is almost always necessary to attach furring strips when you install siding. The pattern in which the strips are applied depends on the type of siding used. No matter which variety you select, the siding must be attached with strong, rust-proof nails that are long enough to extend at least 1 inch into wall studs.

Solid Board Siding

Probably the earliest type of solid board siding was rough logs used to make log cabins. As home building advanced, these logs were replaced by rough boards nailed to the basic house structure. Since then, solid board siding has evolved into a smooth, quality finished product. They are available either in standard lengths or frequently in lengths designed for creating patterns. To make installation easy, they have lapping edges, bevels or interlocking grooves. The cost of this siding varies; the grade, type and number of cuts required determine the cost of the material.

The type of wood for siding is very important. It must be easy to work with, yet free from warp and easy to finish. Two of the best woods for this purpose are redwood and cedar since they possess a natural resistance to decay. Eastern and Western white pine, sugar pine, and Cypress are good woods, but Douglas fir and Western larch and Southern yellow pine are only fair.

When selecting wood for board siding be sure that it is stamped *clear* or *select* and that they are vertical grain boards of heartwood. Such wood has a greater resistance to warp than flat-grain types. The wood can even have some knots as long as they are tight and they are treated before installation. Any board lumber should be certified kiln dried. Wood that isn't properly dried will probably develop problems. Although wood can be treated after installation, it's much more convenient to purchase lumber that is preprimed or treated with a water repellent or sealer to protect it from moisture and make painting easier.

Most board siding is applied horizontally. Some products are best installed vertically. Occasionally products are installed diagonally to produce an interesting effect. Be sure to consult with your supplier for details.

If you have never installed board siding, it might be best to stick to the tongued-and-grooved variety. It offers good protection and the material is easier to align during installation than with some other types.

Once you've made your selection there are a few points to remember. It's best to plan areas around doors and windows before you begin working. Wood in these areas should run continuously without notching and corners should be butted together snugly.

Board Siding Tips

Wood for board siding tends to be soft. This means that it will mar or damage somewhat easily if proper care isn't taken. Make sure that the wood is protected from moisture. Siding should be kept off the ground and covered.

The location of nails used for fastening the siding is important. If nails are driven near the edge, the siding will probably split. If nails must be positioned near the edge, predrill the nail holes. If this isn't practical use blunted nails.

Plywood Siding

One of the easiest types of siding to install is the plywood variety. Because it's laminated (layered) it's extremely strong. The large size of each panel—they are available in 4 by 8 foot and larger dimensions—means that it can be installed rapidly. This type does not have to look boring. There are many finishes and surfaces for plywood siding. There is even a type that's available in long panels of 12 inch to 16 inch widths if you want to combine the strength of plywood with the look of board siding.

How to Install Plywood Siding

Plywood can be applied directly to studs without sheathing. If this is done, be sure that the plywood is at least three-eighths inch thick when the studs are on 16 inch centers. If the studs are spaced wider than 16 inches use plywood that is at least one-half inch thick. The versatility of plywood allows it to be installed over existing products if you have a firm base. You can probably use five-sixteenth inch plywood over such material. However, applying plywood siding over sheathing offers additional strength. Plywood that is five-sixteenth inch thick can also be used for this purpose. Remember, the thicker the plywood the more rigid it is. While it's nice to have this rigidity, thick plywood is also heavier and so is more difficult to work with.

- Carefully plan the type of plywood siding you want to use. Consider the size and finish and whether the product is to be installed over an existing material, sheathing or studs.
- Decide whether the panels will be installed horizontally or vertically.
- Apply the panels. The technique for this differs slightly when the panels are installed horizontally than when installed vertically. When applied directly over studs horizontally, the vertical edge joints should be staggered and the long horizontal edges should be fastened to fire stops or nailing blocks.
- Nail the panels into position as follows: Position nails around the edges of each panel. The nails should be spaced approximately 6 inches apart. Nail each panel to intermediate supports. The nails should be spaced about 12 inches apart.
- The size of the correct nails depends on the thickness of the panels you are installing. Six penny nails should be used with three-eighths inch and one-half inch thicknesses, while 8 penny nails are used with five-eighths inch sheets or when the wood is lapped or beveled.
- Seal all edges with a coat of primer, sealer or water repellent preservative if the wood was not pre-treated. This will help to keep moisture from entering the home and it also protects the siding. However, even when the wood is protected, it will still absorb some moisture which will cause it to expand. Be sure to allow about one-sixteenth inch between panel ends and edge joints so that the wood has room to expand.

Wood Shingles and Wood Shakes

One of the most popular of siding materials, wood shingles and wood shakes are versatile as well. Not only are they used as siding, but they are frequently installed as roofing and for interior decoration of walls and ceilings. They come in a wide assortment of colors, materials, textures and quality. Wooden shingles are graded by a numeric code; number one is made from 100 per cent vertical grain heartwood and is the best. Number four is really designed for interior use. Numbers two and three are intermediate grades.

Wood shingles come in different lengths that you can specify when ordering. Widths are generally random but you can get standardization if you purchase *dimension shingles*. There is one type that is convenient for the do-it-yourselfer. It comes as large panels of shingles that have been bonded to one-half inch plywood sheathing.

If you plan to shingle a side wall you should use a special shingle that is *rebutted and rejointed* with square edges. This is a processed shingle.

Shakes are a special form of shingle that come in three varieties. *Hand split and resawn* shakes are easily recognizable. They are thin at one end and thick at the other and have a face that is rough and a back that is smooth. The second type *taper split*, resemble the hand split and resawn shakes, but both faces are rough. *Straight split* shakes make up the third category. They have a regular thickness and are rough on both sides.

The shake you should use depends on where and how it will be used and the desired overall appearance. Shakes that are used on roofs take a beating from rain, snow, ice and heat. Those used as siding must combat the same elements, but because they are on the sides of a building they are not subjected to the intense conditions a roof must endure.

Installing Shingle and Shake Siding

Although there is a wide assortment of shingles and shakes to choose from they are installed in somewhat the same manner. Two techniques are used. *Coursing* refers to a row of shingles or shakes that are placed side by side. The first technique, *single coursing* calls for one row to overlap the one below it by one-half the length of a shingle or shake. To promote greater exposure of finish shingles or shakes, the *double coursing* technique is used. Here a layer of low quality material is installed and then another layer of high quality is layed over the first. Use the following steps for single course installation.

- Determine the number of required courses and proper shingle or shake exposure, by measuring each wall to be covered from its top to the foundation.
- Subrtact 1 inch to allow for the region directly above the foundation.
- The exposure of the shakes or shingles depends on the size. For shingles the amount of exposure is as follows: 16 inch length—exposure is 7½ inches for a single course and 12 inches for a double course. 18-inch length—exposure is 8½ inches for a single course and 14 inches for a double course. 24 inch length—exposure is 11½ inches for a single course and 16 inches for a double course.
- Shakes require slightly different figures: 18 inch length—exposure is 8½ inches for a single course and 14 inches for a double course. 24 inch length—exposure is 11½ inches for a single course and 20 inches for a double course.
- Take the exposure value for the items you are using and divide it into the number obtained in the second step. This is the number of rows of shingles required. It's a good idea to get some extra shakes or shingles to compensate for breakage and for border areas. Some types of shakes or shingles could be difficult to match exactly at a later date due to variances in materials and finishes.
- Use a story pole (see the section on stairs) and mark the location of the courses on it. It might be necessary to allow for some variance to make sure that butts are aligned with window sashes and eves.
- Transfer the markings on the story pole to the corners, windows and door casings.
- To provide a guide for installation, snap a chalk line between the lowest marks above the foundation.
- Nail 30-pound felt along the studs or over sheathing.
- The spacing requirements for shakes in single courses differ slightly from those of shinges. Note: When you attach shakes in single courses the nails used should penetrate the backing by at least 1 inch. They should be driven slightly above the line for the next course and spaced about 1 inch from each side of the shake.
- Use special rust-proof 4-penny nails.

- Start at the bottom of each wall and install two layers of material. Place nails about 1 inch above the chalk line, at three-quarters of an inch spacings for shingles 8 inches wide or narrower. For wider shingles position an additional nail. Be sure to drive nails carefully. Hammer marks will ruin the appearance of the shake.
- Snap a chalk line between the next two marks over the shingles just installed.
- Temporarily tack a long straight board against the chalk line. This serves as a guide for rapidly aligning the shingles. Be sure that each joint between adjacent shingles. Be sure that each joint between adjacent shingles or shakes is tight. Avoid having the joint between two adjacent shingles or shakes align with a joint between those in the next lower row.
- It is necessary to provide a butting surface for the shingles or shakes at the corners of the wall. This is done by nailing 1½-inch square wood strips to any inside corners.
- Overlap shingles as necessary at outside corners or bevel them to fit snugly against each other. If the corners do not fit together properly, they should be shaved with a utility knife. Special finishing molding can be used in corners.

The technique for positioning shingles in double courses is somewhat similar to the previous steps. Because they are placed over different layers, shingles should be fastened with 9-penny rust-proof nails. Unlike single course methods, the bottom course consists of three layers of shingles. There is an extra undercourse on top of the regular undercourse. The finish course fits on top of this extra layer. Each undercourse shingle is nailed only once. Edges should be spaced one-eighth inch apart. After each layer of undercourse has been fastened across the bottom row, each finish shingle is placed on top and about one-half an inch lower. This is done by driving nails about 2 inches above the shingle butts and about three-quarters of an inch in from the side. Once this is accomplished, drive additional nails at about 4 inch intervals across the shingles using 5-penny rust-proof nails. The butt between adjacent shingles must be tight. These joints should not align with similar joints in the row immediately below.

Double coursing with shakes is somewhat similar. They should be nailed through the butts and along the butt line of the next higher row.

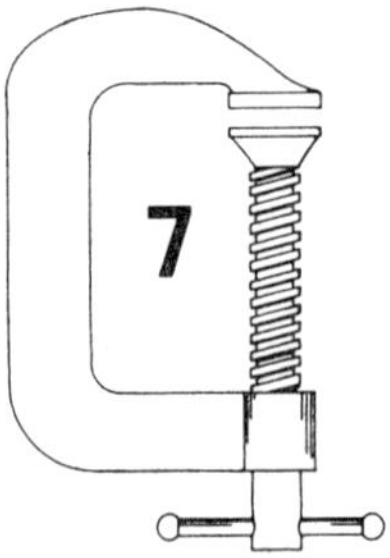

7 Carpentry Projects

Although the use of tools and a knowledge of materials is interesting, you can really demonstrate skill and creativity when constructing projects. While construction techniques are most useful to the homeowner, there is an almost unlimited number of jobs that can be done by any carpentry oriented individual. In most cases, the projects presented here can be completed with basic skills such as measuring, nailing and sanding. The one item that can't be measured is the enjoyment derived from completing each item and the satisifaction of a job well done.

Some of these projects rely on terms or techniques described in other chapters. So that the best results are obtained from your efforts, it's a good idea to briefly review any items you aren't completely sure of before you start working.

PANELING

Although generally thought of as a "home improvement" project, paneling can be installed either in the home or apartment. However, if you install paneling in an apartment some holes will be made in the existing walls. The landlord will probably require that the holes be repaired if the paneling will be removed when you move. There is an extremely wide assortment of paneling to choose from and different ways to install it. However, choosing and installing paneling doesn't require advanced skills. Molding can be used to provide finishing touches or to mask errors or misjudgments; the beauty of the finished project can be judged as work progresses.

There are three basic types of paneling, *solid board, sheets* such as plywood and hardboard and *gypsum wallboard*. The solid board and sheet types come in an extremely wide assortment of colors, woods, finishes and thicknesses. *Tongued-and-grooved boards* are easy to install and add warmth to any decor. Special boards with square edges can be applied vertically or cut and applied to create special patterns. Boards are available in smooth finishes for a polished effect or in rough-milled facing that gives a rustic look. The wood doesn't have to be free of faults. Knotty woods can look beautiful in the proper setting.

Sheet paneling is the type that most frequently comes to mind when the term paneling is mentioned. They come in large sized sheets (usually 4 by 8 feet) and cover a large area quickly. Because they are available in so many finishes—real wood, veneered layers, simulated wood prints, pictured prints to name only a few—there is a panel selection for almost every budget. Most sheet paneling is available on plywood for added strength.

Another type of sheet panel is made from hardboard. It comes either pre-finished—the most popular—or unfinished. Hardboard panels are available with realistic wood grains. Because of the way they are constructed they are generally more durable than "real" wood. But their biggest advantage is low cost. Because they aren't made from whole wood, grain matching and faults are not problems. The result is a product that is less expensive than actual wood.

Gypsum wallboard can also be used to finish a room. It is fast, relatively inexpensive and easy to install. Since it can be painted, it can be used with almost any decor. Installation of wallboard is similar to that of wallboard ceilings discussed in chapter 6.

Paneling Tips

Although the actual method used to install paneling differs with the type of material selected, there are few basic tips you should understand before you begin work. They will speed the job, avoid waste and produce more professional-looking results.

No matter what type of panel material you select, be sure to handle it carefully. Most sheet panels damage easily and the result of mishandling can be a wasted sheet. This will of course, increase the cost of your project.

It's a good idea to store paneling in the room where it will be installed for about a week prior to the actual installation. Panels tend to shrink slightly after they are installed. This storage time permits the wood to attain the same moisture content as the room and thereby reduce post installation shrinkage.

The best results are obtained when the wood is cut as little as possible. Planning the installation before you begin work is very important. A few calculations before the panels are placed will result in a much better looking wall. Make sure that panel edges will align with doors and windows. One frequent mistake made by beginners is to make the final panel too narrow. This is avoided by careful pre-planning.

Try to avoid using molding except around the perimeter of the room or to cover unavoidable cuts. Even though molding hides mistakes, its real purpose is to provide a finished look. The use of too much molding detracts from the final appearance.

Paneling is frequently applied to furring strips, but if the wall is flat and in good condition this probably isn't necessary. It can be applied directly to an existing wall with supply panel adhesive available at lumber yards and building supply centers. Additional holding is supplied by nailing panels into studs. Make sure that the nails are long enough to penetrate at least 1 inch into the studs. Nail holes can be touched-up so they don't show or better yet, special nails can be used that match the panel finish. When you are driving nails, particular care must be taken to avoid damaging the panel surface with the hammer. To reduce marring, try the following trick. Place a thin piece of cardboard over the panel where the nail is to be driven then fasten the nail through the cardboard. Use a nailset to position the nail head below the panel surface. Then pull the cardboard away from the panel. If thin cardboard isn't available, wrap the hammer head with cloth to cushion any contact it makes with the surface.

Whenever walls are not flat or existing walls can't support the paneling directly, install 1 by 4 inch furring strips vertically. If you are using 4 by 8 foot sheets, space the furring strips every 48 inches so that the sheets will be nailed to them around the edges and thereby minimize nail exposure. When horizontal board paneling is installed, position the furring strips vertically at 32 inch intervals. If you use vertical boards, install furring strips horizontally. No matter which direction the furring strips are installed, be sure that they are nailed to the studs inside the wall so that the nails penetrate the studs by at least 1 inch to provide good holding strength.

The easiest way to install paneling over furring strips is to use a thin ribbon of adhesive where the panels meet the furring strips. Extra support can be provided by special nails or by touching up nail heads so that they don't show.

One of the trickiest parts about installing paneling is making the panels at the vertical edge of one wall fit properly against the adjacent wall. Even in new homes, rooms might not be exactly square and walls can be irregular. If the adjacent wall has previously been covered with some irregular material—brick face or stone for example—the matching process becomes more difficult. Of course, molding can be used in corners to eliminate the problem, but results aren't always satisfactory. The problem can be solved with the use of an inexpensive compass that has one pointed leg and a second leg into which a pencil fits.

Matching Panel Edges to Irregular Adjacent Walls

- Determine the edge of the panel.
- Place the panel against the wall where it will be installed. Brace it against the wall so it will not move. Be careful not to mar the finish. The edge of the panel should be about 1 inch from the adjacent wall.
- Hold the compass so that both legs are parallel to the floor. In other words, so that both legs are horizontal.
- Run the pointed leg over the irregular wall in the corner. As the compass is moved, a matching pattern will be traced on the paneling.
- Cut the paneling with a coping saw, by carefully following the traced pattern. The panel edge should fit properly against the adjacent wall.

BUILDING INTO WALLS

One problem with many homes is that no matter the size of the rooms, there never seems to be enough storage space. Building extra rooms, attics or basements can be expensive. Yet there is a way of adding some storage space without taking away living area. Simply build into existing walls. Built-ins can be made to hold books and magazines, small toys, knick-knacks, canned goods and other small items. There is practically no limit to the items that can be combined into wall space. Since walls aren't very thick, bulky items usually can't be stored in this manner. Because there can not be good circulation inside wall units, it's not a good idea to install lighting fixtures inside them.

Wall storage areas need not be hidden. When carefully planned, they provide a handsome addition to a room. For example, shelves built into a bedroom wall can be attractively finished to provide space for night-time reading material, an alarm clock and other objects normally kept near your bed. By constructing them at

Fig. 7-1. Two easy projects are built-in shelves for knick-knacks and the headboard eliminator with added magazine rack.

bed height, the bed can be pushed lengthwise against them. The result is an attractive, functional inside-the-wall cabinet that eliminates the need and cost of a headboard.

For some projects, it might be necessary to actually cut wall framing members, but for many others the technique is quite simple.

While the techniques for opening a wall vary with the scope of the project, the steps presented here can be used in the construction of any shallow cabinet. Typical of the items that can be built using them are medicine chests, in-wall ornament or figuring holders, small book shelves and magazine racks. In fact, you are limited only by your imagination. Figures 7-1 and 7-2 show a few of the many space saving projects that you can construct.

— Closely examine the wall to be opened. Avoid opening a wall near plumbing, electrical switches and outlets and heat ducts. If you don't know what's behind the wall covering, tap the wall sharply and listen. Objects contained in the wall can often be discovered in this way. If listening doesn't reveal anything, carefully tap a small nail

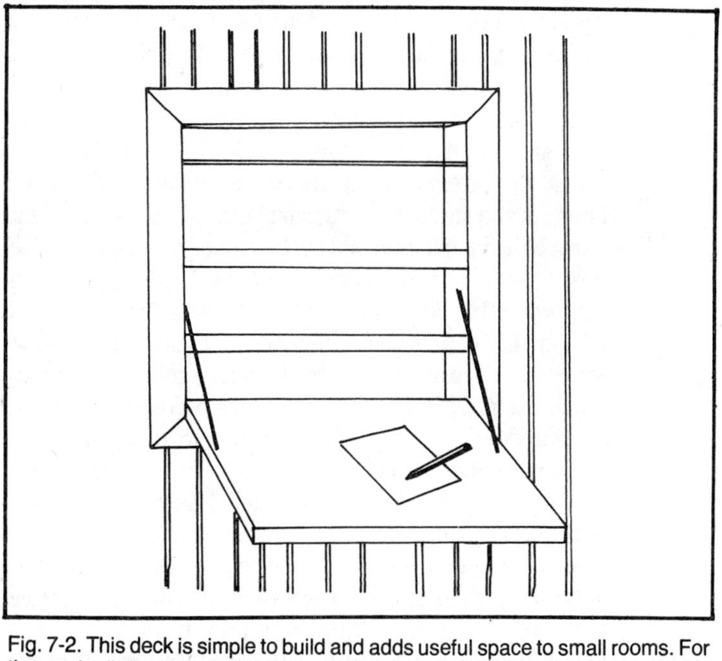

Fig. 7-2. This deck is simple to build and adds useful space to small rooms. For those who live in apartments a modified version can be hung on a wall so that the wall need not be opened.

into the wall. Stop at the first sign of increased resistance since this means that something is hidden.

— Mark the size of the opening on the wall. Depending on your plans, there might be a few extra inches required around the entire opening for molding. Be sure to allow for it.

— Depending on the project, it might be necessary to cut wall members. Nailing can be done more easily if extra room is provided around the opening. It's recommended that the wall covering be opened a few inches wider than the finished plans call for so that this extra space is available.

— Knock a small hole in the center of the region you marked in the second step. Use a flashlight and check inside the wall for plumbing, electrical wiring or heating items. If any are present it's best to fill in the hole and try a different location.

— Use a small saw, utility knife or hammer and chisel to cut away the wall covering.

— If the project requires some depth inside the wall, it will be necessary to cut the studs. Projects such as built-in medicine chests frequently don't need this extra space—so this step can be skipped if it does not apply. Studs are generally 16 inches apart on centers so if the required width is narrower than this, your best bet is to shift the opening so that stud cutting is avoided. If you must cut into studs, carefully mark where the cuts are to be made. Then cut the studs. If studs must be cut, local codes might require the installation of 2 by 4 inch headers and sills. Be sure to allow sufficient space for them. When the opening is between 16 and 48 inches wide, 4 by 4 inch headers and sills should be provided rather than 2 by 4 inch sizes and wider openings might require stronger lumber. Be sure to check the requirement for your area before you cut the studs and leave enough room for their installation. Note: If it's necessary to cut studs to make an opening wider than 4 feet, the ceiling support could be weakened. It's advisable to provide temporary support until sills and headers are positioned. This can be done by placing a 2 by 4 flat against the ceiling a few feet away from the wall that is opened. Use a few 2 by 4s to prop it up.

— Install the header by cutting the wood so that it is the same length as the distance between 2 full (uncut) studs.

Then drive nails through the full studs into the header wood. Fasten other nails through the surface of the header and into each cut stud.

— Install the sill in a manner similar to the sixth step.
— To complete the frame for the opening—position cripples, the lumber running between the sill and the lumber running between the sill and the header on both sides, by nailing them to the studs.
— Once the frame has been completed, the next step depends on the particular project. Generally, some sort of backing is positioned and then any shelves or cabinet work.
— Fill any gaps around the opening with spackling compound. After it dries use fine sandpaper over spackled areas so that they are as smooth as the adjacent wall regions. If necessary, paint the spackled portions to match the surrounding wall; because of difficulties in exactly matching some shades, it may be necessary to paint the entire wall.
— Molding around the opened area provides a professionally finished appearance.

BUILDING A SPACE SAVING KNICK-KNACK SHELF

Figure 7-1 shows how knick-knack or small book shelves can be built into a wall. These shelves will provide attractive storage space without taking up valuable living area.

— Decide on the best location for the shelves and determine the approximate size of the finished shelving.
— Follow the first eight previously covered in steps Building Into Walls.
— Construct a simple frame from 1 by 4 inch wood as shown in Fig. 7-3. The dimensions of the frame depend on the size of the opening that has been made. It should just fit into the completed wall opening.
— Cut a piece of one-quarter inch plywood so that it is the same size as the outside dimensions of the frame in the third step. See Fig. 7-4.
— Nail the plywood to the back of the frame.
— Measure the distance between the side members of the frame. Mark this distance on the wood from which the shelves are to be made (1 by 4 inch lumber if the frame is 1 by 4).

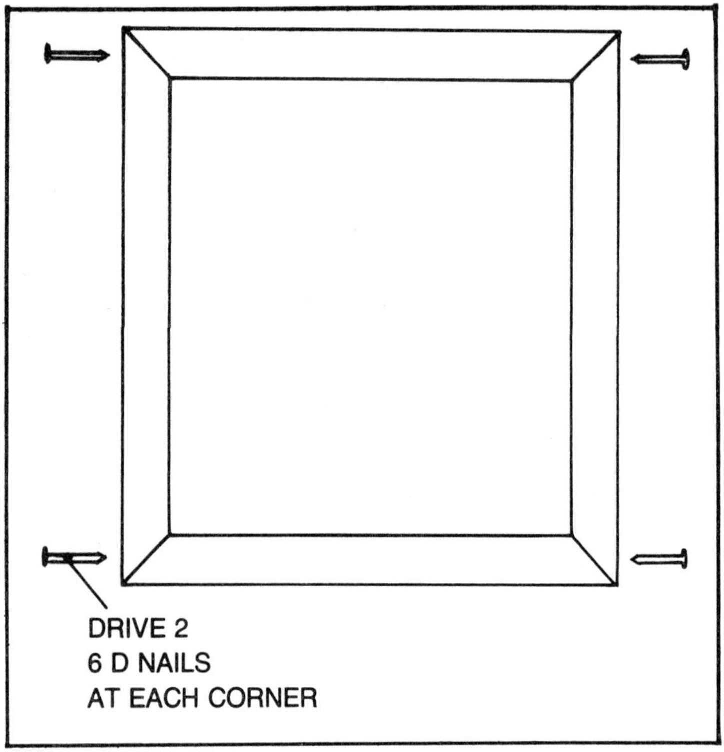

Fig. 7-3. A square frame is used to hold shelves or compartments. It consists of 2 long legs and 2 shorter legs. The longer lengths are installed horizontally for strength. The sides are fastened to each other with 6 penny nails, two at each corner.

— Cut the shelves to the marked length.
— Determine the desired spacing between shelves and mark the sides of the frame where the shelves are to be placed.
— Nail the shelves to the frame by driving 1 inch nails through the frame into the shelves. Use two nails for each end of every shelf.
— Slide the frame into position inside the wall opening and nail it to the wall frame.
— Set any nails that show and cover them with a dab of wood filler.
— Lightly sand the frame and shelves with fine sandpaper. The shelves and frame can be stained, painted or finished as desired.
— Follow steps ten and eleven of Building Into Walls.

When these steps are followed, the entire frame and shelves will be flush with the wall cover. If deeper shelves are desired, use wider wood. If this is done, the shelves will extend out from the wall.

HOW TO BUILD A CUPBOARD

No matter how big a kitchen is, most homemakers will tell you that there isn't enough storage space. By applying the same idea as the knick-knack shelves, a handy cupboard for small items can be built into a kitchen wall. Figure 7-5 illustrates the construction of the in-wall cupboard.

- Determine what items the cupboard will contain. Heavy cans require greater support than spice jars, so consider using heavier support lumber than called for in the following steps if heavy items are to be stored.
- Follow the instructions given for Building Space Saving Knick-Knack Shelves.
- Cut a piece of decorative plywood the same size as the plywood backing.
- Attach the decorative plywood to the jamb with hinges to form a door.

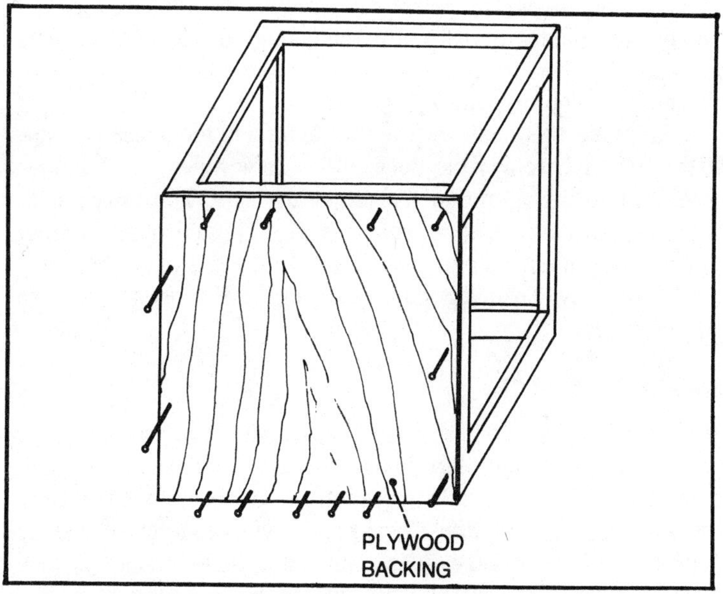

Fig. 7-4. Cut a piece of plywood to the outside dimensions of the frame to provide a backing. Use 1 inch nails to attach the plywood to the frame.

— Lumber yards and building supply departments of do-it-yourself stores carry a wide selection of door latches and handles. Select a combination that bests complements your kitchen decor and attach them to the door and frame.

HOW TO BUILD A MEDICINE CHEST

Once you've mastered the steps for building a cupboard, constructing a medicine chest for the bathroom is easy—simply follow the same directions. For added convenience, attach a mirror to the outside of the door.

HOW TO BUILD A SPACE SAVING DESK

Another variation on the cupboard is a desk that opens out from the wall. This is particularly useful in a small child's room or sewing room. Small shelves can be used to hold water paints, pens, pencils, crayons and small toys or different types of threads, needles and sewing machine attachments. While most sewing machines would probably be too heavy to be supported by the desk top shown in Fig. 7-2, the surface could be used for tracing patterns or for notes and sketches.

The same style desk could be used in a small foyer to hold a telephone and important telephone numbers, in a dining room to hold glasses and bar utensils or in the garage to hold shop manuals and small tools and gauges. You can even use it in the work area to plan and develop carpentry projects.

To make this desk, follow the directions for making a cupboard. When fastening the door to the frame, use sturdy hinges screwed into the sill rather than along the side. To provide additional support, attach stranded wire to each side of the frame. Connect the other ends of these wires to the desk top about one third of the way in from the end. Use decorative screws for a more attractive look.

BOOKSHELF PROJECTS

Whether you live in a house or apartment, you probably don't have enough shelf space for books, knick-knacks or stereo equipment. The solution—handy bookshelves—are easy to make. Both the novice and the professional can obtain beautiful results by applying basic carpentry skills. Units can be as simple as pine shelving attached to walls with pre-made brackets or as complicated as a fine piece of furniture with hand-crafted cabinets. The bookshelves described here require different degrees of skill to complete, but don't be limited to them. Part of the fun is to design your own and successfully accomplish them. Both the novice and the experienced individual will find building these bookshelves easy if you remember the following rules:

- **Plan.** Many errors can be found and corrected through the careful planning of the project before you begin. Ob-

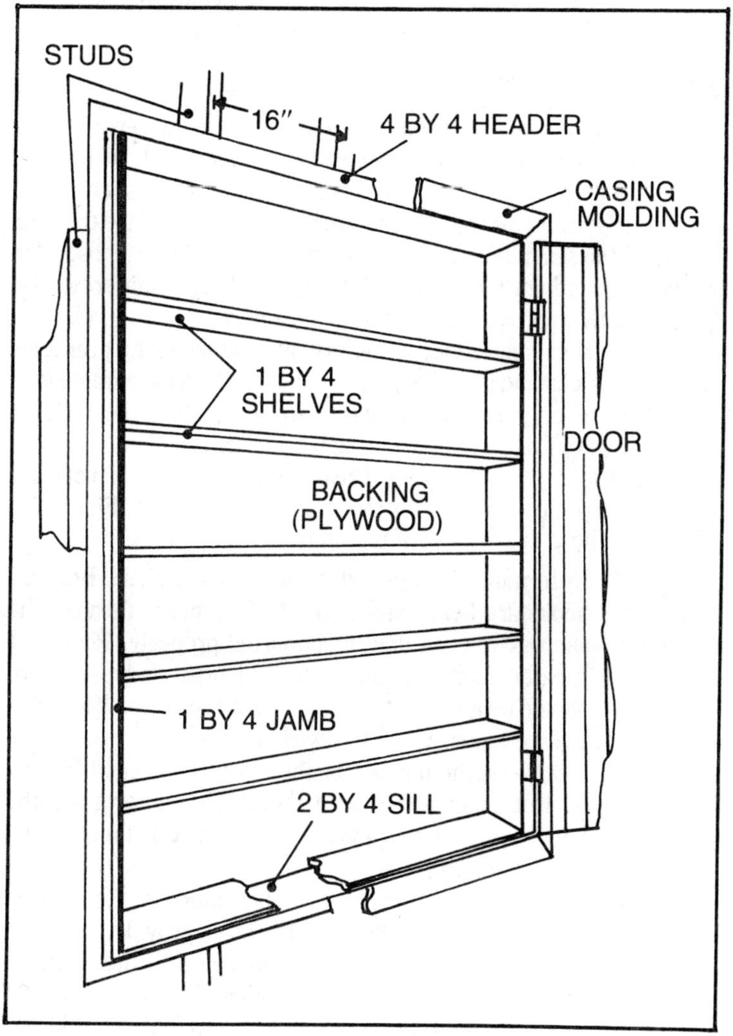

Fig. 7-5. The same techniques can be applied to the construction of a kitchen cupboard or bathroom medicine chest.

tain the materials required before work is started to avoid annoying interruptions.

- **Stick to basics**. The final result can be simple or fancy but the techniques of good woodworking should always be applied.
- **Experiment**. Sticking to basics shouldn't stunt your creativity. Don't be afraid to make mistakes because that is part of gaining experience. The challenge of correcting mistakes can add to the rewards of a finished project.

BOOKSHELF PROJECTS

The first bookshelf project—simple wallshelves—is ideal for the beginner. Because of its simple functional design it can be used to complement almost any room arrangement. The shelves are particularly useful for visually breaking up long walls. Since they can be easily disassembled and stored, they are ideal for those who move frequently or for apartment dwellers who aren't interested in permanently improving the landlord's property.

You will use basic measuring, sawing and fastening principles and the work can be completed fairly quickly. Because a frame isn't required, the only materials required are the shelves themselves, metal brackets and supports available in many styles at lumber yards and do-it-yourself stores. The relatively low price of materials makes this project suitable for those on a strict budget. Figure 7-6 illustrates how this wall shelf unit is made.

- Determine the desired length of each shelf. Because books are heavy and most shelf lumber is flexible, the shelves will sag unless supported properly. Plan for a greater number of supports for longer shelves. There should be a support for every 48 inches of shelf length.
- Locate studs within the wall.
- Decide on the number of shelves you will build and the desired spacing between them. This determines the length of the shelf bracket supports. Cut them to the proper length.
- Attach the first shelf bracket support to the wall by positioning screws through the holes provided. Be sure that the screws extend far enough into the studs to provide good holding power; otherwise the shelves will pull away from the wall under the weight of books.
- Align each shelf bracket support with the first one. This is important because the shelves will slant. If the bracket

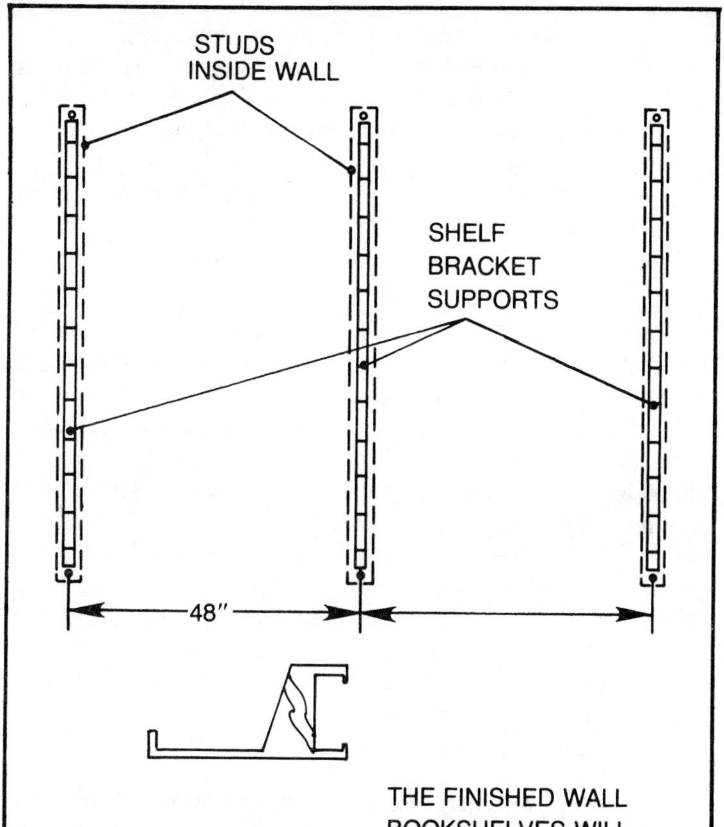

Fig. 7-6. One of the simplest bookshelves to make consists of brackets, supports and shelving lumber. If you redecorate or move, these shelves are easily disassembled.

are not level. When they are aligned properly, attach each one to the wall as in the previous step.

- Place shelf brackets on the supports by engaging them in notches built into the supports. Count down the same number of notches on each support so that the brackets align with each other.
- Mark the shelf wood with the proper length and then cut to size.
- Lay the shelves in the brackets. They are now ready for use.

Part of the utility of these shelves comes from the number of variations that are possible with the same basic project. Essentially the same bookshelves can be installed in different rooms and yet they will look dissimilar. For example, the shelves just described were unfinished. They could be stained to match room furniture or painted to match or contrast walls. Molding can be placed around the shelves for an added touch. Molding too can be plain, stained or painted. The shelves can be made from fancy woods. Even if they are made from simple shelf lumber, they can be made decorative by scalloping the edges or beveling them. Try tracing the bracket pattern onto thin decorative lumber and cut the wood twice for each bracket. Then glue it to the bracket to form a "wood sandwich". Use your imagination for other ideas.

PLYWOOD BOOKCASE

If you own many large heavy books or if you don't want to attach anything to a wall, a bookcase made from plywood might be just for you. It's more difficult to construct than simple wallshelves, but can be accomplished by someone with only a little experience. The trick to this project is measuring each cut carefully and then making the cuts accurately. Figure 7-7 illustrates how the components are made and fit together. Although specific dimensions are indicated, they can be changed to fit your needs.

— Take a piece of three-quarter inch plywood and cut it to the desired length and height. Figure 7-7 represents a piece that is 80 inches long by 16 inches wide.

— Starting at one end of the plywood lay off the desired shelf spacing.

— At each spacing marking draw a rectangle 7⅝ inches long and three-quarters of an inch wide. These will serve as notches into which the shelves will fit. Be sure that the last notch marked is at the opposite end of the plywood.

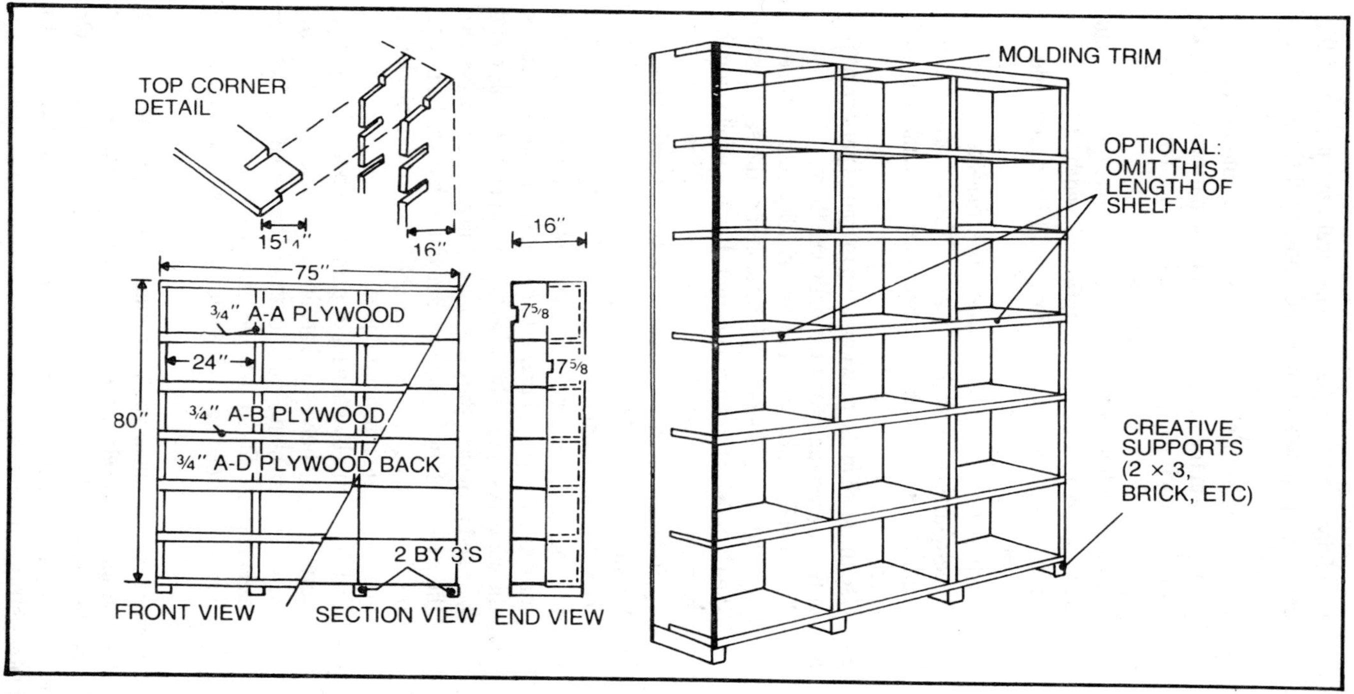

Fig. 7-7. A plywood bookcase is simple to build and forms an attractive complement to any room.

- Carefully cut the notches. The final appearance of the bookcase depends in part upon how well these notches are made.
- For each 2 feet of shelf space desired, cut another piece of plywood as in the first step. Trace the pattern cut in step four onto each piece and carefully cut it. You have now completed the vertical supports.
- For each shelf desired, including the top and bottom, cut plywood to the desired length and width. Figure 7-7 represents pieces 75 inches long and 15¼ inches wide as shown.
- Starting at one end of the one shelf, mark a 7⅝ inch by three-quarters of an inch notch. Mark similar notches separated by 24 inch intervals. Then carefully cut each notch.
- Use the shelf in the previous step to trace a pattern on the other shelves and then carefully cut them.
- To provide extra strength and rigidity, cut a sheet of plywood so that its outer dimensions are the same as the shelf length by the support height. If you are following the diagrams, the size will be 80 inches by 75 inches.
- Fit together all the shelves and supports and nail on the backing.
- You will notice that the bottom shelf has no support. Here is another opportunity to be creative. The drawings call for 2 inch by 3 inch lumber to be cut to the same width as the supports—16 inches. However, other materials could be used providing they are sufficiently strong to support the weight of the books or other times that will be placed on the shelves. For example, bricks, cinder blocks or squared stones might be used.
- Finish the unit by staining or painting. Then add molding between the shelves and around the outside of the book-case. Use finishing nails to attach the molding. Set the nails and use fill so that the nails don't show.
- Finish the molding so it either matches the shelves or contrasts with them.

For a different touch, you could make one of the shelves only one-third as long as the others. Plants, small lamps, statues or other decorative items can be used to break up the pattern of books. Or you could attach a hinged door over the large compartments to create a storage cabinet.

FRAME BOOKSHELVES

A different type of bookshelf that adds a great deal of variety to a room is made from 2 by 4 inch framing lumber. Because of the strength of this wood, plywood backing isn't required in most cases. The finished product is supposed to have a look rough. Accidental hammer marks or small splits add to the appearance of the shelves, so this is another good project for gaining experience.

- Cut 2 by 4-inch lumber into 4 sections, each 4 feet long. These strips will form the base and top of the bookcase.
- Cut 2 by 4-inch lumber into 4 sections, each 40 inches long. These will be used as vertical supports.
- Cut 2 by 4-inch lumber into 8 sections, each 1 foot long. These will be used as horizontal side supports.
- Line up a vertical support with one end of a base section so that the wide side of the base section is vertical. Then drive two 8-penny nails through the base section into the vertical section.
- Repeat the previous step for the other end of the base section.
- Repeat steps 4 and 5 for the second base section.
- Line up one of the 1 foot sections with the end of a base section and then nail it into position with two 8-penny nails driven through the base section into the 1 foot piece. Then attach the other base section to it in a similar manner.
- Repeat the previous step for the other ends of the base sections.
- Because 2 by 4 lumber might not be cut exactly to those dimensions, measure the distance marked A in Fig. 7-8. Cut four sections of 2 by 4-inch lumber to length A.
- From the top of the base sections, measure up 11 inches on each vertical support. Using 8-penny nails, attach two of the sections cut in the previous step to the vertical supports so that they are at this height.
- Fasten two of the 1-foot strips between the vertical supports, one at each side. Again use 8-penny nails.
- Measure up 11 inches from the top of the supports just positioned and attach the other two A-length sections and two 1-foot pieces as in the two previous steps.
- Align a 4-foot section on top of two vertical supports and nail it into place with two 8-penny nails at each side. Then repeat this for the remaining 4-foot piece.

- Install the remaining two 1-foot pieces at the top of the frame at each side. The frame is now complete.
- Cut shelves to fit the frame, including one across the top of the frame. Here is another opportunity to be creative. The shelves can be fashioned from simple plywood or from wide rough planking. Almost any material of the proper size can be used providing it's strong enough to support the weight of books or the other items that you will keep on the shelves.

FURNITURE PROJECTS

Many carpenters have an urge to either build or refinish furniture. Depending on the item, furniture can be relatively simple to construct or require the skills that can only be gained through years of experience. For the beginner, furniture projects offer the opportunity to develop skills before you attempt major construction jobs. You can derive considerable pleasure from constructing difficult pieces and developing your skills. But whether difficult or easy to make, furniture building provides items that can be shown with pride.

One such project is refinishing an old dresser or bureau. New bureaus made completely of wood are difficult to obtain, and when they are available they are very expensive. Sometimes old dressers can be obtained very cheaply at garage sales or perhaps from a neighbor who is discarding one. If you are fortunate enough to obtain such a bureau and the wood is still sound, even if it is cracked or damaged, you will be able to refinish it so that it looks new. The cost of refinishing is relatively small. However, depending on the shape of the piece, considerable effort could be involved so don't expect results overnight.

REFINISHING PROJECTS

The specific steps to follow for refinishing any piece of furniture vary. Different woods are employed in the original construction and they might be split or dried out with age. Finishes vary with the style, piece and manufacturer. The item could also have been previously refinished or painted. However the following steps apply in general and can be adapted to almost any situation.

Before you begin, it's a good idea to obtain a good power sander—both coarse and fine sandpaper and a paint scraper. The power sander will save hours or labor and will come in handy in a variety of projects. The scraper is a great help in removing old paint

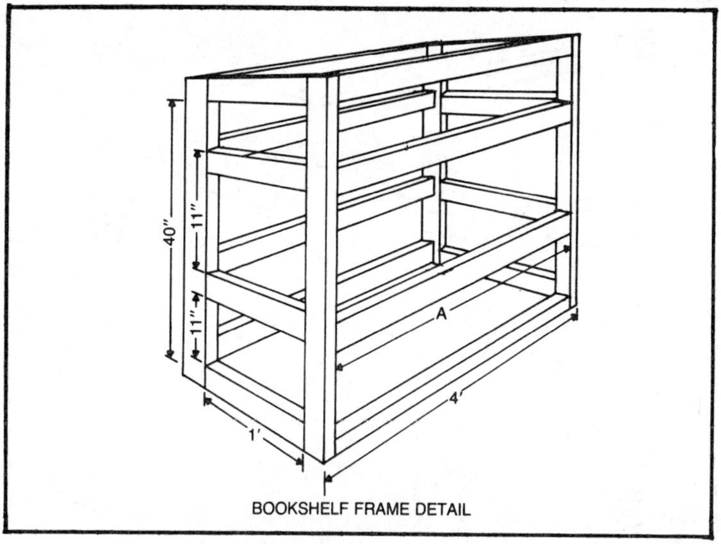

Fig. 7-8. A bookshelf made from framing lumber is easy to construct and lets the beginner practice construction techniques.

or thick finishes. There are several chemical finish removers available. Follow the directions on the package to remove paint and other finishes.

How to Refinish a Dresser

- Remove all drawers from the dresser and examine them for loose joints and splits in the interior wood. If shelf lining has been previously glued in place, pull it off. Use a scraper if necessary.
- Remove all drawer knobs or handles. If they are in bad condition or you don't care for their style, discard them.
- Take an old sponge or cloth and wet it slightly. Lightly wash the entire bureau inside and out. This will let you see the condition of the wood and any loose items that need tightening.
- Old or misused furniture often has loose items such as drawer tracks or joints. Depending on the part, nail or glue it in place.
- If you plan on using a chemical finish-remover, apply it at this time. Let it work for a while and then scrape it off in accordance with package directions. Be sure that the work area is well ventillated. Breathing vapors from stripping chemicals can be harmful.

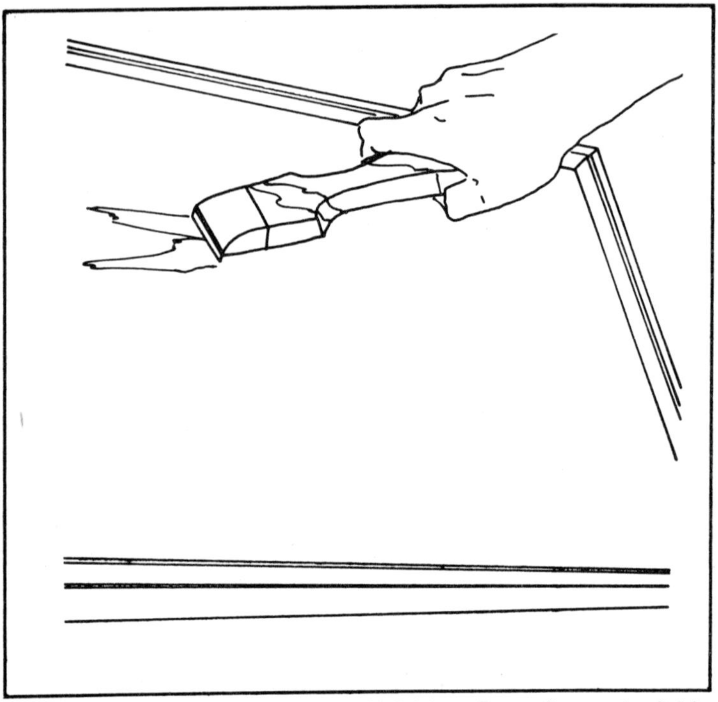

Fig. 7-9. A scraper helps to remove old finishes. Depending on the finish, considerable effort may be required. The old finish should be completely removed before a new finish is applied.

- Use the scraper to remove as much of the old finish as possible (Fig. 7-9). The old finish should be completely stripped before a new finish is applied.
- Any finish that couldn't be stripped by scraping can be removed by sanding. If most of the previous finish has been removed, a good power finishing sander can be used (Fig. 7-10). Of course, both hands should be positioned on the tool before the power is switched on. The chemicals in finishes sometimes "bleed" into the wood so it might be necessary to sand quite a bit to remove this. In such cases, sand across the grain of the wood with a medium or coarse sandpaper and then sand again in the direction of the grain. Don't be concerned about scratches made by the sandpaper as long as they are not deep.
- If you don't plan on using the existing drawer knobs or handles or if any are missing and can't be matched,

select those that you plan to install. If the holes in the drawer don't match the handles you will have to fill the holes. Use a putty or spackling knife and wood filler. If the wood is very thick, fill each hole about half way. Allow the filler to dry and then fill the remainder of the holes. Fill all holes slightly higher than the surrounding surfaces and feather the filler outward from the holes. See Fig. 7-11A and B.

- Examine each part for splits, cracks and damage to the wood. Use wood filler to fill all irregularities or holes. As in the previous step, fill all such items slightly higher than surrounding surfaces and feather the filler outward.
- Again using the power sander in the direction of the grain, sand the entire surface with fine sandpaper. Use long strokes and press slightly harder in the area of filled holes and marks until the entire surface is level and smooth. Any scratches from previous sanding must be removed.
- Mark the location of new holes for the handles you want to install. They should be positioned on the drawers so that the handles are about 6 inches in from the drawer

Fig. 7-10. A power sander is best for removing small areas of old finish from furniture. Sand across the grain to remove stubborn finishes or stains and then finish along the grain.

edges. However, the distances can be placed to suit your tastes.

- Drill holes with a diameter slightly larger than that of the screws provided with the handles. Always drill holes from the outside toward the inside of the drawers. Depending on the handle design, it might be necessary to countersink screw holes inside the drawers, so that items placed in the drawers will not catch on screw heads.
- Using fine sandpaper, hand sand along the grain of the wood until all pencil marks are removed. If you are planning to stain the wood, sanding is extremely important. Lightly sand all edges and then sand the entire wood surface again. Don't forget to do the inside of each piece to remove dirt and provide a "new" appearance.
- Vacuum or brush away all sawdust. Then take a sponge that is slightly damp and run it along the grain of the entire bureau. This will remove any small traces of sawdust that could affect the final finish. Be sure that the sponge is only damp because too much water damages wood.
- Stain or paint the wood. See the next chapter for techniques and suggestions. Apply any special finish protectors.
- Align each drawer handle with the holes you made and insert and tighten screws.

DESIGNING AND BUILDING TABLES

One of the best and most practical projects for the beginning carpenter is designing and building a table. Although fancy tables require considerable carpentry skill, simple ones consist of little more than a large piece of wood supported by four legs. Even though such tables are simple in concept, the variety of leg styles and types of wood and finishes allows you to express your own creativity.

— Select the materials you will use to build the table. There is a wide assortment of pre-made legs, both wood and metal, available at lumber yards and in do-it-yourself centers. Or, you can try designing your own styles.

— Make a rough sketch of the table showing the table top, its dimensions and where the legs are to be positioned. If you want to make the legs yourself, draw a separate sketch of them.

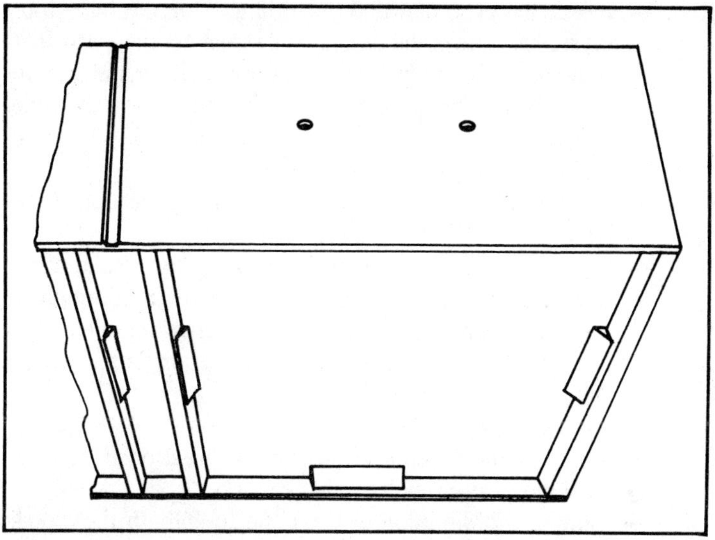

Fig. 7-11A. Holes such as these might not correspond to the handles you plan to use. They should be filled.

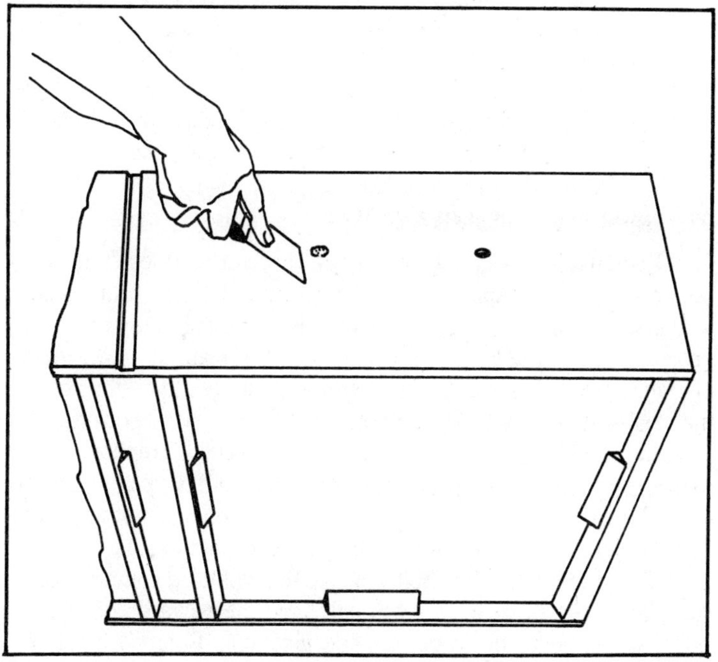

Fig. 7-11B. Use wood filler to handle holes. Deep holes should be filled in two stages.

— Select the type of wood you will use for the table top. Plywood is one of the easiest and strongest materials that could be used, but fancier woods or plastics might also be considered. Choose a material that you feel comfortable handling.
— Cut the table top to the exact size.
— On the back side of the table top mark the location of the legs.
— If you have designed your own table legs, make them now. It is important that all the legs be the same length or the table will be unsteady. Be sure to provide a means for attaching the legs to the table. Brackets or clamps will do the job.
— Attach the legs to the table top.
— Sand any unfinished wood across the direction of the grain to remove any rough spots or marks.
— Sand in the direction of the grain to smooth the wood, eliminate any scratches and prepare it for finishing.
— Finish the wood by staining or painting. See the next chapter for finishing techniques and hints. If you prefer, there are several types of vinyl and plastic materials that can be placed over the wood. They generally are self-adhesive or special glues can be used. Some individuals don't like the layered look around the edges of plywood. If you do not cover the plywood, you can hide the edges behind decorative molding.

DESIGNING AND BUILDING A CLOSET

Even if a home has many closets, there often doesn't seem to be enough closet space. If you understand basic framing and wallboard techniques, adding a closet isn't difficult. As with many projects, one of the most important aspects is planning. Considerable work can be saved by selecting the best location. The cost of materials can be reduced if use is made of existing walls and offsets.

Constructing a closet involves the building techniques of measuring, cutting and assembling framing lumber. You must also know how to use a level. Considerable experience is not required. This project is excellent for developing carpentry skills.

The closet described below is one that you plan yourself. Once the plans have been made, you will have the rewards of actually building a project from your own specifications. Figure 7-12 shows the general framework of a closet that is fitted to a corner of a room. Use it as a guide in developing your own plans.

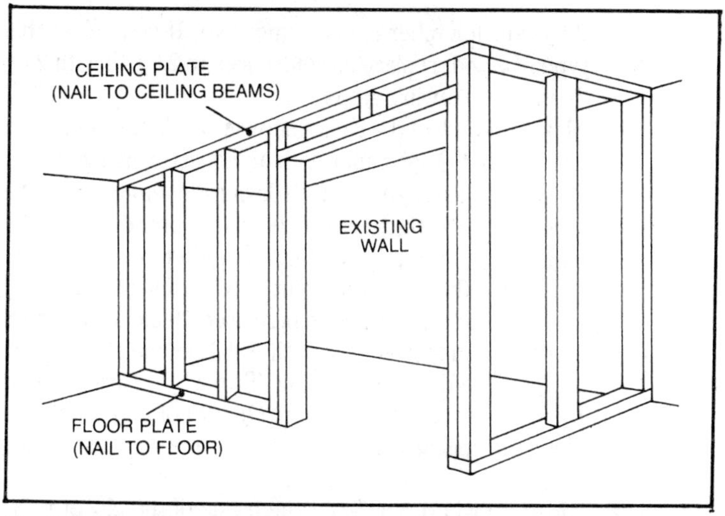

Fig. 7-12. Closets aren't difficult to make once basic construction techniques are mastered. Using existing walls helps to reduce work and lower cost.

How to Build a Closet

- Determine where you want to place the closet. Use existing walls to your advantage. Look for offsets that can be enclosed or build the closet into a corner so that existing walls serve as the back and side.
- Decide if you want shelves built into the closet and allow for their spacing. If the closet will be used for clothing storage make sure that it will have enough depth so that the clothing will have room to bang easily. Plan on installing a rod across the length of the closet to support hangars.
- Closets should have doors. Select the type to be used—folding, sliding or conventionally opening doors.
- Depending on the location of the closet you might want to install a light. This light can be either a built-in or plug-in electrical type or you can select a battery operated one. Refer to TAB book No. 913, *The Complete Handbook of Electrical & House Wiring* for information on installing lights. Battery lights do not require special wiring.
- Select the type of lumber to be used in construction—2 by 4-inch framing wood is a good choice because it's strong. If you like, 2 by 3-inch wood could be used. The

2 by 3-inch lumber is less expensive. If you select this type of wood, replace all references to 2 by 4s with 2 by 3s in the following steps.

- Make a sketch of the closet showing all dimensions. You will find that graph paper with one-quarter-inch boxes is useful for planning a layout. In laying out lengths, be sure to allow for the thickness of the plasterboard that will be used over the framing lumber. The sketch will help you determine the proper amount of wood and plasterboard needed. Allow for studs to be placed 16 inches apart on centers, starting against each wall. Double studs go on each side of the door opening. Two 2 by 4s are used for the header over the doorway. One set of 2 by 4s is used along the ceiling and another serves as a sole plate along the floor.
- Measure where you want the frame of the closet to be positioned and mark the outline on the ceiling, walls and floor. Masking tape is often useful in marking floor dimensions.
- Once the dimensions have been marked on the floor it is easy to transfer lines to the wall. Simply place a 2 by 4 against the wall in the appropriate spot so that one edge of the wood aligns with the markings on the floor. Check to make sure that the wood is truly vertical by using a level. Then run a pencil along the edge of the lumber from floor to ceiling. Next, use a carpenters square to extend the lines onto the ceiling so that the pattern on the floor is duplicated.
- If the existing wall has ceiling or floor molding in the closet region, carefully remove them. For certain types of molding, it might be necessary to remove a larger strip than the closet dimensions call for. This molding can be cut to the correct size and replaced after the closet is completed.
- Locate ceiling beams by tapping the ceiling. A change in sound from hollow to a solid dull thump indicates that you have located a beam. Remember that beams generally run along the width of a room rather than its length.
- Mark the location of the beams found inside the area of the closet.
- It's important that the pattern on the ceiling match the one on the floor. Before you start work check this by

dropping a plumb bob from the outer corner of the ceiling pattern. It should meet the corner marked on the floor.

- Install 2 by 4-inch lumber along the sides of the closet. To do this, first measure and cut the wood to the correct size. If the closet floor is made of wood or has a wooden subfloor, nail one piece to the floor and the other to the ceiling along the marked lines. If the floor is made of concrete, you should use the proper hardware and drill suitable holes. Since the closet walls bear no building weight, you might be able to avoid this procedure by nailing the base wood firmly to the newly installed studs. Then install another 2 × 4 along the length of the closet along the floor as a brace. Use 10-penny nails. Nails used to attach wood to the ceiling should be driven into the beams.
- Measure and cut 2 × 4s to fit along the other markings on the ceiling and floor. The piece running along the ceiling should be uninterrupted. When measuring the wood, remember to allow space for double studs on each side of the door opening. Keep in mind that the door and its frame will fit into the door opening and a small amount of clearance will be required.
- Position the corners of the closet. This is done by measuring 2 ×4s to fit between the floor plate and ceiling wood. Nail the 2 × 4s into place against the wall by driving nails into studs that are inside the wall.
- Install the outer corner of the closet. This is actually a single 2 × 4 or if extra strength is desired, two 2 ×4s. Nail the corner wood to the floor and ceiling plates.
- Cut as many 2 × 4s as needed to fill the closet wall between the 2 × 4s against the wall and the corner and doorway. Space them 16 inches apart on centers. Drive 10-penny nails at an angle to fasten them to the floor and ceiling plates.
- Measure, cut and nail two 2 × 4s to the floor and ceiling plates on each side of the door opening. Allow clearance for the door to swing.
- Consult your plans for the height of the door to be used. If a carpet will extend into the closet, allow room for the door to clear it. Also allow room for the thickness of the door sill plus three-quarter of an inch for the head jamb and an additional one-quarter inch clearance. Measure

and cut two 2 × 4s to fit across the top of the doorway. Position them on the studs but don't install them yet.

- Two additional studs must be installed on either side of the door opening. These go between the door saddle (yet to be positioned) and the bottom of the header. Measure and cut the wood. Mark the existing studs with the proper position of this wood. After marking the studs, nail the additional lumber into position against the studs.
- Nail the headers you previously positioned on top of the studs.
- Short pieces of 2 × 4 should be positioned between the ceiling plate and the header. If the door opening is a standard width, only three pieces are necessary. Install them over each side of the header and over the center of the door opening. If you have planned for a wide door, place sections 16 inches apart on centers if possible—otherwise space them about evenly. All pieces should be toe-nailed to the header and ceiling plates.
- You are now ready to install the surface of the closet wall. If your plans call for plasterboard, measure out from an existing wall and plan to install the plasterboard horizontally. Be sure to start and end on a stud. Cut the plasterboard to size and nail it into position against the studs you have installed. Use 4-penny nails if you are using three-eighths-inch plasterboard or 5-penny nails for the one-half inch type. If you prefer, special purpose nails are available for this job.
- Set all nails. Tape and use spackling compound to cover all joints and nail heads. Feather the compound outward, let it dry and then sand until smooth. The procedure is the same as the steps required for installing walls and ceilings. Refer to those sections.
- Measure 1 by 6-inch wood to be used as side jambs. Cut the lumber and nail it into place using 6-penny finishing nails.
- The head jamb goes into position under the header. Measure 1 by 6-inch wood and cut it to the appropriate size. Use 6-penny finishing nails to attach it to the header.
- You are now ready to attach molding to cover the joint between the side jambs and plasterboard. If you want to

be a little fancy, cut the top corner at an angle—preferably in a miter box. Cut the molding to the proper size. Use 4-penny finishing nails through thin molding into the jamb and 7-penny finishing nails through the thick part of the molding into the plasterboard and studs.

- Measure and cut trim for the side molding and then nail it into position to the head jamb and studs with 6-penny finishing nails.
- The procedure to follow at this point depends on the type of door you plan to use. Generally, door stops should be installed. Use one-half by 1-inch wood that is mitered at the corners and temporarily nail them in place. Hang the door and then nail the stops firmly into position. The door should be against the stops when closed with a one-thirty second of an inch clearance.
- Install any shelves or clothes rods inside the closet. If the closet is longer than 4 feet, use a center bracket so that the rod will not sag under the weight of heavy clothes.
- Inspect the walls of the closet for any cracks and fill with spackling compound if necessary. Then let dry and sand smooth.
- Paint the closet walls and either paint or stain the trim.

DESIGNING AND CONSTRUCTING PICTURE FRAMES

An ideal carpentry project to help pass time on a rainy or dreary day is the design and construction of picture frames. One advantage of this type of project is that the frames that are made can be simple or fancy depending on your level of experience. Simple or fancy, there are really only two types of frames—those that consist of sections fitted together (Fig. 7-13) into which the picture is placed and a solid sheet that supports the picture itself.

Solid Picture Frames

This type of frame consists of one solid piece of wood. The picture is glued or otherwise attached to it and a protective see-through cover is placed over the entire surface. They are easy to make and are quite attractive when completed. Best results are achieved when pictures on plain paper are used. Don't use your best pictures until you get the hang of the project. Practice with magazine pictures.

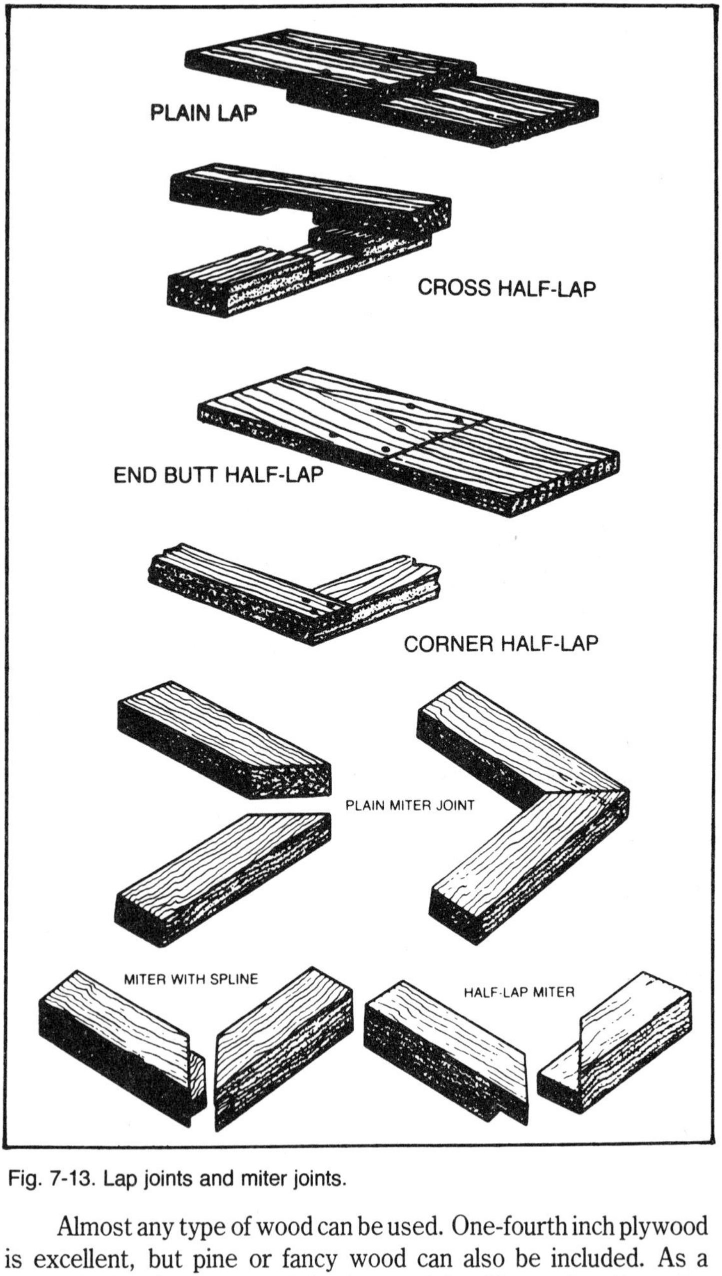

Fig. 7-13. Lap joints and miter joints.

Almost any type of wood can be used. One-fourth inch plywood is excellent, but pine or fancy wood can also be included. As a general rule, choose a wood that has a subdued but noticeable grain. This type of grain offsets almost any picture without detracting attention from it. Figure 7-14 shows the basics of a solid picture frame.

- Select the wood to be used. It can be plain or fancy. Wood with slight flaws often adds to the visual appeal of the final product. Rough surfaces can give unusual and attractive results. For your first frameless frame its best to use a fairly smooth wood.
- Measure the picture size and mark the wood so that each side is at least 1 inch bigger than the picture.
- Carefully cut the wood. Try to avoid splintering—particularly when you are cutting across the grain.
- Use fine sandpaper and hand sand in the direction of the grain. Slightly round each edge of the wood. When sanding the sides of plywood, start at the edge and work toward the middle.
- Remove all sawdust from the wood. A clean surface is important or the results will be bubbly.
- Center the picture on the wood to get an idea of how the finished frame will look. Place a light pencil mark on the wood at the corners of the picture.

Fig. 7-14. The final appearance of a solid frame. Best results are achieved with wood that has a subdued but noticeable grain.

- Remove the picture and apply clear shellac or plastic to the wood. Be careful to avoid bubbles.
- Allow sufficient time for the finish to get slightly tacky. Align the picture with the light pencil you had placed at the corners. Press the picture onto the wood and carefully remove any air pockets by running the edge of your hand over the picture. The picture must be flat against the wood.
- Let the shellac dry. Then apply a second coat of shellac or clear plastic over the entire picture and wood. Again be very careful to avoid bubbles. Allow sufficient time for the finish to thoroughly dry.
- Attach picture hangers to the back of the wood and hang the picture.

There are some interesting variations you can accomplish using your own creativity. For example, use a match to lightly singe the edges of the picture before placing it on the wood. Of course don't set the picture on fire. Smother the sparks before they fully ignite and keep shellac or any flammable substance away from the flame. The result is an interesting aged affect ideal for pictures of old maps and similar subjects.

Another attractive idea is to emphasize the edges of the wood by noticeably bevelling them. Then apply gold leaf paint. The result is a formal look.

Wood Frames

Attractive and unusual picture frames can be made from the same wood used in construction. The wood can either be finished roughly or smoothly. This is a project suitable for both experienced carpenters or beginners. Accidental scratches and blemishes add a rustic quality to the frame.

- Measure the size of the picture to be framed.
- Measure and cut two pieces of 1 by 2-inch wood so that each piece is 3 inches longer than the *length* of the picture. Measure and cut two additional pieces of 1 by 2-inch wood so that each piece is 3 inches longer than the *width* of the picture.
- Place the four pieces of wood on a flat surface so that they form a rectangle. Arrange them so the long pieces overlap the adjacent shorter pieces.
- Place the picture over the wood sections and adjust the

wood so that the opening between them is slightly smaller than the picture itself.

- Use a square to make certain that the wood has right angles. Then carefully check the opening once again.
- Nail the wood together using two nails at the places where the pieces overlap. For extra strength, toe the nails slightly toward each other.
- The final appearance is now up to you. Turn the frame so that the nails don't show. Sand in the direction of the grain with fine sandpaper and then stain the frame. The result will be a smooth finished appearance. As an alternative, place the frame so that the nail heads face you. Lightly sand with medium sandpaper and stain the wood. Lightly sand again for an interesting rustic effect. Or be creative and experiment until you've achieved your own special look.
- Tape the picture to the wood. If additional rigidity is needed cut a piece of thin plywood to size and nail it to

Fig. 7-15. Frames made from construction wood are inexpensive and easy to make.

the back of the frame behind the picture. Make sure the nails don't damage the picture. Figure 7-15 illustrates how the wood frame will look when completed.

For an interesting variation, paint the frame a color to match the scheme of the room. Let the paint dry and then sand in the direction of the grain. The paint will give a highlighted effect that is both artistic and unusual.

Another finish that you can try is staining. A dark, flat stain results in a rugged look. Rough spots and marks on the wood emphasize the rough appearance and highlight the picture.

BASIC PINE CABINETS

Cabinets, both plain and fancy, appear to be little more than boxes. They look like they should be easy to make. Yet it often takes several years for an individual to become a skilled cabinet maker. The artisan must learn the characteristics of many kinds of wood, fully master woodworking techniques and understand the use of tools. The relationship of one wood to another is a factor and working with veneers must be mastered. Wood finishing skills must become second nature. But don't be discouraged. There are cabinets that the average carpenter can construct without years of practice.

Cabinets can be built into walls, hung on walls, or stand free. Since built-ins have already been discussed, I will discuss some of the hanging and stand alone types.

The basic cabinet made from a soft wood such as pine is an ideal project for the beginner. The skills of measuring and truing edges will be developed and soft wood is easier to work with than hardwood varieties. This cabinet is designed to hang on a wall and it can be built in about two days. Figure 7-16 shows different views of the cabinet. Dimensions will vary with your needs.

- Decide on the dimensions of the cabinet. Until you gain some experience don't plan on a large heavy unit. The basic cabinet will consist of four pieces of pine shelving. Additional lumber will be used for the inside shelf and doors.
- Although the experienced carpenter can plan fancy joints, these plans call for simple connections of the sides. Be sure to allow for the width of the wood when figuring overall sizes.
- Cut the lumber to size. Beginners should measure and cut one piece for the longest side, place it against the

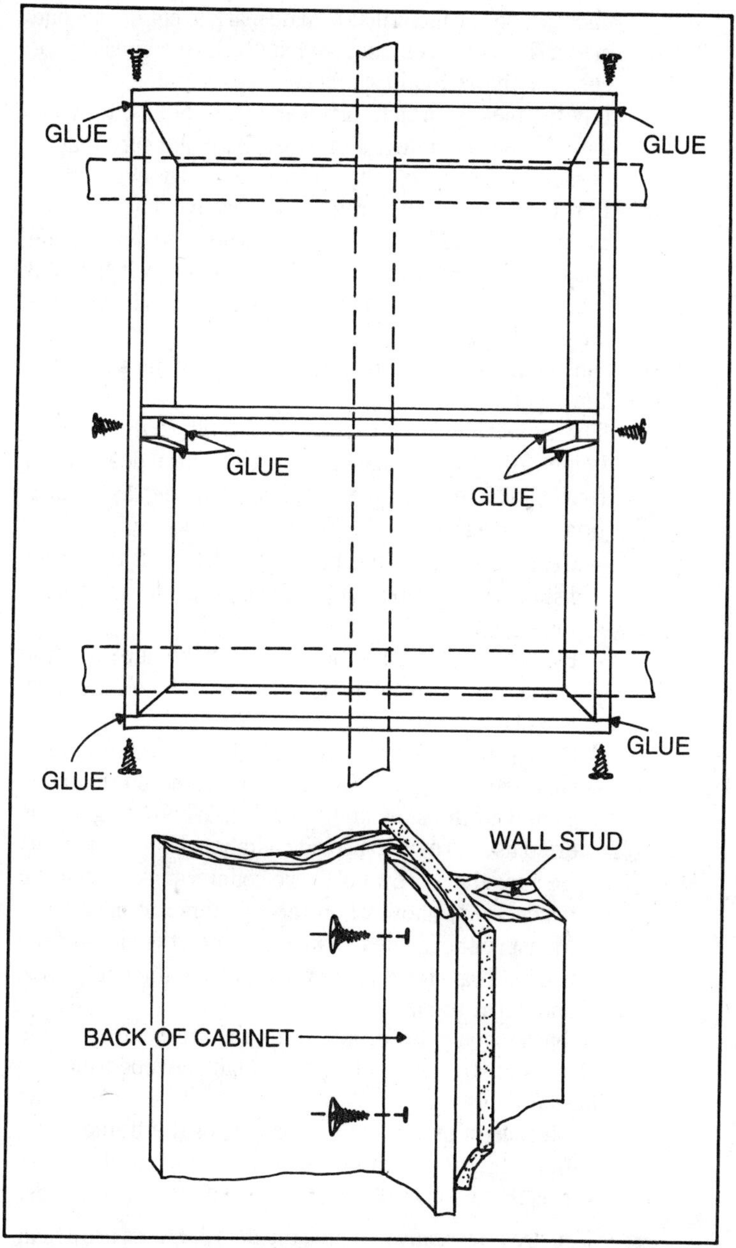

Fig. 7-16. A simple type of wall cabinet. The top view shows the construction of the cabinet and the bottom illustrates how it is hung on the wall. Doors and hardware are not shown. To attach some types of door styles it might be necessary to attach additional wood as shown by the short dotted lines.

uncut wood and using it as a guide mark to cut the second piece. This ensures uniform length. Use a similar process for the second two pieces.

- Lay the pieces on their sides and glue them with wood glue. When clamping them together, make sure that all corners are square. Then let the glue dry overnight.
- For extra strength, drive nails through each corner toeing them slightly toward each other. Be sure to use nails long enough to penetrate at least one-half inch into the second piece of wood. Make sure the nails you use have wide heads.
- Set the nails so that the heads are below the surface of the wood.
- The next step is to position the shelf at the desired height. Although an adjustable shelf is often desirable, a fixed shelf is easier for the beginner to install. To position the shelf:
  —Measure up from the bottom of the shelf the exact distance where you want to place the shelf. Then draw a line.
  —Place a level against the line and make sure that the line is truly horizontal. If the line is not level, the shelf will slant.
  —Repeat the previous two steps for the other side.
  —Glue small strips of wood to the frame so that they align with the lines on both sides. When the glue has dried use screws to provide additional holding power. Be sure that screw holes are countersunk so that the heads won't show when the cabinets are finished.
  —Measure the inside distance between the sides of the frame. Transfer this measurement to the shelf wood and cut it to size.
- A back should be attached to the frame to provide increased strength and rigidity. Thin plywood usually is best for this job.
  —Measure the outside dimensions of the frame.
  —Mark the wood with these lengths and cut it to size.
  —Nail the wood to the frame at about 12-inch intervals.
- You are now ready to install the doors. Doors are available pre-cut or you can make them yourself. They are an important part of the cabinet and their appearance plays a large role in how the final job looks. If you want to make

them yourself, choose a wood that matches the frame. Decide if you want to use one or two doors and cut them to size. Select from a wide assortment of door handles, hinges and latches and position the doors. Make any necessary screw holes but don't attach the doors until the cabinet has been finished.

- Fill in all holes and make sure that nails can't be seen. Then sand the frame in the direction of the wood. If the doors are not pre-finished sand them at this time.
- Paint or stain the cabinets and the doors if they haven't been pre-finished. Allow sufficient time for the wood to dry and then attach the doors.
- The cabinet is now ready for hanging. Locate studs on the wall where you will position the cabinet. Mark the wall for screw holes. Holding the cabinet near the wall, mark the back with the corresponding location for screw holes. Take care in marking because the holes in the cabinet must align with the holes in the wall.
- Make the holes in the wall and cabinet and mount the cabinet on the wall.
- Insert the shelf on the supports.

Some types of double doors might require additional supports for the door latches. These doors are generally designed to overlap the front of the cabinet. If you plan on using this type of door, cut two pieces of wood to the inside length of the frame and another piece that will connect them as shown by the short dashed lines in Fig. 7-15. Glue and nail them into position within the frame so that they are flush with the other pieces of the frame. If the shelf is the full width of the cabinet it will have to be notched to accommodate the additional vertical piece.

When you are building this cabinet or any piece of furniture, all pieces must be square or the results won't be the best. If there are spaces at the joints, the cabinet won't be sturdy. Small spaces can be filled with wood filler and sanded prior to finishing.

There are many modifications to the basic hanging cabinet design that you can incorporate for variety. Instead of installing a support for a single shelf, metal tracks could be used. They have slots for supports that allow the height of the shelf to be adjusted or permit the use of more than one shelf. They are available at many do-it-yourself centers.

Special attachments can also be added. For example, holders for mugs or cups are available in many gift departments. Or, an

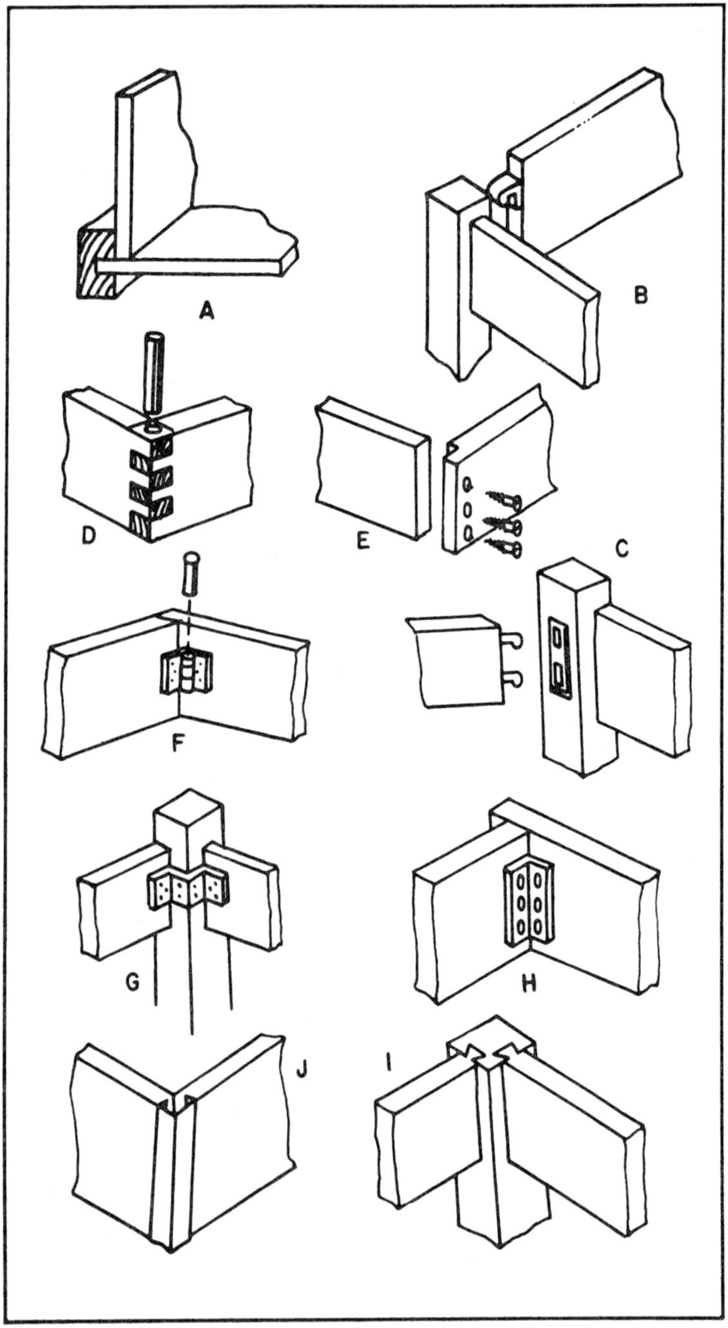

Fig. 7-17. Various frame corner treatments.

unusual effect can be created by the use of special lights. If the cabinet is to be used in a kitchen, dispensers for foil or waxed paper or paper towels can be added. Decorative molding can be used to highlight such items.

The same basic cabinet can be modified for use as a free standing unit. Many do-it-yourself centers and lumberyards sell furniture legs. They come in a wide assortment of styles and finishes. They generally either screw on the base of the furniture or can be forced-fitted. To modify the wall cabinet into a free standing unit:

- Cut four blocks of wood approximately 3-inches on a side.
- Drill a hole the same size as the part of the leg that either has a screw thread or will be inserted into the cabinet.
- This next step differs slightly depending on whether the legs are the force-fit or screw-on type. If the force-fit model is used, simply sand the hole in each block until it is just wider than the leg. Then force the legs into each hole. If the screw-on type is used, you must provide a screw thread. One way to do this is to glue a metal washer to each block. The washers should have holes the same diameter as the parts of the legs that are to join the cabinet. Once the washers are fixed the legs can simply be screwed into place.
- Glue the blocks onto the bottom of the cabinet about 1 inch in from each corner. For extra strength, drive nails through the blocks into the bottom of the cabinet. The nails should not pass through the bottom wood.
- Finish the leg wood to match the rest of the cabinet.

There is another type of leg you can use—although it isn't as readily available as the screw-on or force-fit types. It has a bracket attached to it and can simply be screwed into place, eliminating the need for blocks. Some kitchen tables have the special brackets that can accommodate metal legs. If you can find some of these brackets, they can be adapted for cabinet legs and eliminate the need for blocks.

There are many other ways to make standing cabinets. Since part of the fun of carpentry is creating new ideas, don't be afraid to design your own project. See Fig. 7-17 for more ways to construct joints.

8 How to Finish Wood

Wood is a natural material with its own built in beauty. Even when untreated it has a personality and texture. The idea behind finishing wood is to develop the full richness it contains. Wood finishing is easy and the results usually are very satisfying. There are many different ways to treat and finish wood. The one for you to choose depends on the desired effect. Some factors you should consider are whether the wood will be used indoors or out, how it will be used and the type of wood. See Tables 8-1 and 8-2.

RULES FOR PREPARING WOOD FOR FINISHING

No matter what type of wood you are using, there are three basic rules to follow when you are preparing to finish wood. If these rules are followed the finish will produce the greatest benefits.

Start with good wood. Choosing the proper wood at the beginning of a project will result in a better looking job when it is completed. Since there is less chance of shrinkage or warping use kiln-dried wood. Select wood that is free from gouges and damage. Unless they are needed to produce a special effect, try to avoid wood with knots. When knots are present they should be firmly in place. Stay away from wood with fungus or mildew. They are difficult to bleach or cover and an uneven finish will result even if you use more than one coat.

Sand. The importance of sanding to obtain professional appearance can't be over-emphasized. Start with rough sanding to remove high spots or marks in the wood. Use a power sander to

Table 8-1. Exterior paint applications.

TYPE OF FINISH		RECOMMENDED FOR	APPLICATION
PAINT	Acrylic Latex	Medium Density Overlaid, sanded, or 303 textured plywood.	Apply recommended non-staining primer plus two finish coats.
	Oil—Alkyd	Medium Density Overlaid only.	Apply with zinc-free primer, using two or three coats (including primer).
STAIN	Semi-transparent or penetrating stain	Unsanded plywood (303 textured plywood).	One or two-coat systems as recommended by the manufacturer.
	Opaque or highly pigmented stain	Unsanded plywood (303 textured plywood).	One or two-coat systems as recommended by the manufacturer.

Recent government restrictions on the use of lead and mercury compounds in finishes have resulted in some uncertainty regarding the performance of new paints and stains. In many cases, long-term performance information on the substitutes for lead and mercury is lacking, and results of tests in progress will not be available for some time. American Plywood Association finishing recommendations are based on experience with formulations made prior to the current restrictions. Consult supplier for interim recommendations.

speed the job and a rough or medium sandpaper to ease the effort. Then sand again with fine sandpaper to remove scratches and nicks. But even then you're not through. Final preparation for finishing calls for sanding again. This time use a fine sandpaper and avoid using the machine. Wrap the sandpaper around a block of smooth wood. This will help to protect your hands. Final sanding should be with the grain. Avoid rounding edges.

Avoid dust. Dust prevents the bonding of any finish to the wood and the results can look bad. You should always finish wood in a dust-free environment. Since the previous rule calls for sanding and sanding causes dust, it must be removed. Use a vacuum cleaner to remove large amounts of dust from the room an from the wood. Be careful not to scratch the wood or you will have to sand again. Then rub the work with a clean rag.

Remember, the most important part of finishing wood is right up front *before* a finish is applied. The more care spent in the preparation of the wood the better the final job will be. This applies if the wood is new or old, hard or soft.

Once the wood is prepared, the next step depends on the project and the finish you want to apply.

PRIMERS

The purpose of primers is to protect the wood from moisture or decay and to prime the surface for painting. The various products for protecting the wood that you use outdoors have already been discussed, but there is also a special primer that's used with paints. It is especially important to use a primer where swelling of the wood due to moisture could cause a problem, with items such as doors and widows. Pre-painting primers are generally brushed on. Don't be discouraged if the results don't appear attractive. Remember it will be covered by paint and the final job will look better. A paint supplier should be able to help you select a suitable primer for the best results for your project.

In addition to the types of primers that are intended for exterior use, there are those that are used indoors. Generally these are used on plaster or wallboard. These surfaces are porous and unless specially treated often don't absorb paint evenly. One type of primer that is frequently used has a *latex* base—a mixture of various chemicals with water. Because they are based with water, cleanup is easy. However, they generally take overnight to dry properly. Because they don't raise the fibers on the paper covering, they are usually the best primers to use on wallboard.

Table 8-2.Interior paint applications.

INTERIOR APPLICATIONS

TYPE OF FINISH*		RECOMMENDED FOR	APPLICATIONS
PAINT	Oil—flat paint, semi-gloss or gloss enamel	Medium Density Overlaid, regular plywood, striated, embossed.	Apply over stain-resistant primer recommended by manufacturer of topcoat.
	Latex emulsion	Regular and textured plywood, Medium Density Overlaid.	Apply over stain-resistant primer recommended by manufacturer of topcoat.
	Textured paints	Regular plywood, Medium Density Overlaid, Texture 1-11.	Use stain-resistant primer recommended by manufacturer of topcoat.
STAINS, SEALERS		Regular plywood, textured.	Apply stains with companion sealer or over sealer separately applied. Follow with satin varnish o lacquer for increased durability. Where sealer alone is desired, use two coats.

*Use only lead-free finishes

Alkyd primers are good for use under both oil base paints and latex paints. However, they aren't recommended for wallboard since they tend to raise the fibers on the paper covering. If alkyd primers are used on wallboard, an extra sanding step is required before the finish paint is applied.

Other types of primers are designed for special uses. Some are used under an enamel based paint to produce a smooth hard finish while others are formulated to be used as an undercoat for plywood or clear wood. There are even some for use with masonry.

If you paint metal, protect the metal with primers that bond. The most familiar paint primer is red lead, but zinc chromate dries faster than red lead and can be used with aluminum.

TYPES OF PAINT

Anyone who goes into a large paint store will soon realize that there is much more to paint than just its color. There are paints for special finishing jobs, new walls, old walls, special effects, metal and new wood. They are available in gloss (shiny), flat (not shiny) and semi-gloss finishes. There are even some paints that give a sandy or stucco appearance. But no matter which finish you select, paints can be divided into five categories.

Oil Based Paints. These contain a pigment (coloring agent) suspended in oil. They require the use of a solvent such as turpentine or mineral spirits to make them the desired consistency and for cleaning applicators and paint brushes when the job is completed. They are slow drying and give off an instantly identifiable "paint" odor. Although they can be used indoors, they are most often used as house paints or for exterior use.

Alkyd Paints. Their name is derived from a manufactured resin. They are designed for interior use and for interior and exterior enamels. Alkyds are also found in some varnishes. Alkyd paints are easy to apply and dry faster than paints with an oil base. Because they are easy to wash and form a tough coating, alkyds are good in enamels. This type of paint is almost odorless. However, it can be toxic and it should only be applied when there is adequate ventilation. Alkyd paints are thinned with mineral spirits which are also required for cleanup.

Emulsion Paints. This type of paint is made by mixing various chemicals with water instead of with solvents. They generally come in two varieties, *linseed oil* and *latex*. Linseed oil paints are easy to apply, often cover with only one coat and can be used on damp areas. However, when you want to cover bare wood a linseed oil primer should be used first. Latex is the most popular paint. It can be used either indoors or outdoors, is quick drying, covers well, is almost odorless and it can be applied in damp areas. A big advantage is that it can be applied in damp areas. A big advantage is that it can be touched up very easily. New spots tend to blend with previously applied paint. It is best not to apply latex paint to bare wood. Use a latex primer first. A mistake sometimes made by inexperienced painters is to apply latex paint in wet weather. Although latex paint can be applied to cover damp areas, don't use it when it will be exposed to rain within several hours. Whenever possible, it should be kept dry for several weeks after it has been applied.

Water Thinned Paints. In addition to linseed oil and latex paints, there are other water thinned paints. Examples are calcimine and casein paints, whitewash and masonry products.

Catalytic Coatings. The fifth type of paint, called catalytic coatings, contain epoxy and polyurethane resins. While conventional paints rely on evaporation to dry, catalytic paints work through chemical reactions and they produce the toughest finish available for home use. They can be applied to wood, masonry, fiberglass, porcelain or non-porous materials and they are good for marine use. Epoxy paints come in two containers. One container holds the pigment and the other holds the hardener. Once mixed the paint must either be used or discarded. There are disadvantages to this type of paint. Because of irritating fumes they must be applied only in well ventilated areas. They can't be applied over other types of paint and they can be difficult to work with. Although they will still protect, they sometimes chalk when exposed to weather.

Besides being categorized by composition, paints are also grouped according to the location where they will be used and the items they are intended to protect and beautify. While using a paint intended for a specific purpose for non-recommended uses might work, best results are generally achieved by following manufacturer's recommendations. For example, exterior paints are intended for use outdoors. Interior paints, designed for indoor use, may work but are not developed to withstand the rigors of sun, wind, and rain. Floor paints are formulated to withstand the abrasion of feet; other paints will fail. Marine paints are made to survive moisture, salt and chemicals that are found in rivers, lakes, and other waterways.

Painting Tips. In order to obtain the best results, you should prepare the paint before it is used. One important aspect of this preparation is shaking. New cans of paint that you will be using within a few days after purchase should always be placed in the dealer's paint mixing machine. This machine helps make the ingredients uniform throughout the entire can of paint. When the paint is applied it will have a more professional look. Whether or not a mixing machine is used, the paint should be stirred prior to its use. A clean stick or a paint paddle will do. Paint paddles are frequently supplied free by paint dealers. For thorough mixing, professional painters recommend pouring part of the contents into another clean can and then transferring the contents back and forth between the cans several times until complete blending has occurred. Be sure not to stir clear finishes. Products such as varnishes shouldn't be stirred. Doing so creates bubbles which will be transferred to the work.

WOOD STAINS

One of the most beautiful ways to bring out the warmth and individual characteristics of wood is to stain it. Stains allow you to highlight wood features or to change the color without hiding the texture and grain. Stains aren't used over primers. They are absorbed into the wood. It is this characteristic that separates stains from paints—paints work by covering a surface.

Types of Stains

Stains can be divided into categories. The first category is *oil stains*. They have been used for many years to finish paneling an furniture. They have recently achieved a measure of popularity for exterior use as well. Oil stains can be clear or opaque. The clear variety is intended to change the color of the wood and isn't used to mask grain, texture or imperfections in the wood itself. The opaque type is formulated to partially hide the wood.

Alcohol stains are convenient to use because they penetrate the wood and dry rapidly. However, it's this same feature that makes them difficult to use since they sometimes dry faster than expected. It's best to apply the alcohol stain to small areas at a time and then wipe the surface shortly afterward with a clean rag.

Water stains are also available but they are more difficult to work with than oil stains. Some experience in the application of stains is usually required before the best results can be achieved. Until this experience is acquired leave this type of stain to the professionals.

Stains are often used in conjunction with *varnish*, a special clear finish that's thinned with solvent. Generally they are moisture and wear resistant and protect the finish. They are available in different levels of gloss and there are different varieties for special purposes. Once a stain has been applied it should be allowed to dry thoroughly. Then varnish can be applied. In an attempt to reduce the double effort of both staining and varnishing, there are *varnish stains* available. These are actually a combination of stain and varnish. However, in wood finishing you generally get results proportionate to the effort that's applied. Although labor is saved by using varnish stains, the result might not be as attractive as would be obtained by separate applications and the result frequently isn't as durable.

Staining Tips

Unless a special effect is desired, stains should be applied only to surfaces that have been sanded smooth and cleaned of dust. Stains should be applied with either a brush or a cloth pad and then wiped off with a clean cloth after a short interval. Best results are obtained by applying stains to small areas at a time.

The end grain of the wood will absorb more stain than other areas. Unless care is taken, the final effect won't match the rest of the wood. Use only small amounts of stain on the ends. One trick that might be helpful is to apply a small amount of mineral spirits to the ends just before you apply the stain.

A frequent error made by beginners is to choose a stain that is too dark. Remember that if the result is lighter than anticipated additional applications can be made. However, if the result is darker than desired it can be lightened only by bleaching or sanding.

PROTECTING WOOD

Wood is a strong, durable and attractive natural substance. And like any natural item it will age. As wood becomes older, it loses moisture and can shrink or crack. Many products have been developed to protect wood, while at the same time exploiting its full beauty. The correct method for you to use depends on the purpose the wood serves, its location and any previous attempts to treat the wood.

Shellac is a protective finish that has been widely used for many years. It's a clear finish that is made with natural resins in alcohol. Not only is it used as a base for many other finishes but it is also frequently used over finishes on such items as quality furniture. It stops sappy areas from streaking and makes a good finish for wood floors. Shellac is convenient since it dries relatively quickly. About

15 minutes is all that is required for it to be dry to your touch. It is ready for sanding and a second coat after about two hours. Be careful when you purchase shellac. If it has been sitting on a shelf for several months it might not dry properly when used. Look for a date on the can.

Shellac comes in two varieties, orange and white. The orange type has a color to it when dry and should really be used under opaque finishes. It's also less expensive than the white kind. The term "white" is actually misleading. Although there are shellacs that have white pigment, they are usually used as a sealer or for a final finish. White shellac is actually almost colorless.

The best method for the non-professional to apply shellac is with a brush. Allow the shellac to dry slightly. Then, while it is still wet use the tip of the brush to lightly remove any bubbles. Because the shellac is almost colorless it is easy to miss a few spots. If this happens don't try to touch up with fresh shellac once the product has begun to set. Let it dry and then apply another coat. Best results are obtained when fine sandpaper is used to lightly sand over one coat after it dries and before the next one is applied. Apply several thin coats of shellac and sand after each layer rather than painting one or two thick coats of shellac.

Lacquer is another fast drying finish. It yields a hard glossy finish and is frequently used on furniture. It's available both in white and pigmented varieties and is designed for use with a solvent called *lacquer thinner*. It can be used over shellac if you desire and it makes a good base for paints. Lacquer does have a tendency to remove paint so be careful where you use it.

Another product that has become popular in recent years is *polyurethane*. It is a durable coating for both finished and unfinished wood. It forms a protective coat for floors, paneling, furniture, trim and it can even be used on bar tops, It's applied by brushing uniformly outo the item you want to protect. Light sanding is recommended between coats. If one coat is permitted to dry more than 24 hours before the next one is brushed on, then sanding is required before the additional coat can be applied—otherwise it won't adhere properly. Polyurethane coatings are available either in high gloss or satin finishes.

For many years, oils have been used to protect fine wood products. There are many oils for wood protection and some wood craftsmen develop their own formulas. Specially formulated oil finishes provide a soft appearance and excellent protection from heat and grease. Application and touch-up is easy. Simply pour some oil onto a clean soft cloth and rub it into the wood surface. Let the surface become slightly tacky and then buff. For an increased shine, repeat after 24 hours.

Because of the wide number of wax products available and because of the quantity of advertising promoting wax products, many people use wax finishes. Wax will protect the wood and many products also have a cleaning agent to remove marks and dust. If you want to use wax on wood, use a fine quality past wax for the first coat. Rub it into the wood and then buff according to the directions on the can to get the best results. Additional coats or regular treatments could be either liquid or spray varieties since they are easier to use.

HOW TO REPAIR WOOD

Treating wood and finishing it properly slows the deterioration process of wood but it doesn't eliminate it. Time and normal useage combine to cause wear. Wood that does show signs of wear can often be repaired.

As wood dries over a long period of time, joints that once fit perfectly might no longer do so. Such joints can be strengthened by metal hardware, called *mending plates*, which are metal strips with holes for screws. To install them, place the strip along a flat surface so that it crosses the joint and fasten the screws. In just a few moments you can easily have additional holding power. Another version of the mending plate is the *tee plate* which resembles the letter T. It is used to strengthen the joint created when one piece of wood supports another. The *flat corner iron* serves the same purpose in corners where a mitered joint has been made.

There is also special hardware that actually supports wood that has been joined. *Corner braces* and *chair leg braces* are two such supports. Corner braces resemble mending plates that have been bent in the middle, while chair leg braces are squared to fit into corners but also have a rounded edge for bearing the extra weight of someone sitting in a chair. Other reinforcements and supports are available for heavy duty use. Check with your local hardware store or lumberyard for the brace that best meets your specific needs.

Sometimes wood will be damaged from use or in the construction of a project. If the marred areas aren't too extensive, they can be repaired. Holes in wood can be filled in with wood putty. It's usually available where wood is sold. Using a putty knife and filler, fill the hole slightly higher than the surrounding surface. Let the material dry and then sand it smooth and level. If the hole is deep, don't try to fill it in all at once. Do a little at a time until the hole is slightly overfilled and then sand the surface level. Accidental gouges and small cracks can also be filled in this manner.

Although wood putties are available in different shades to match various woods, it is sometimes important to match the wood extremely closely. One technique for doing this is to sand the wood in a place where it normally won't show and then mix the sawdust you get with animal glue. Snce the sawdust is the same shade as the wood being repaired, the final result usually isn't noticeable—particularly when the damaged area isn't very large.

Furniture, cabinets and other items made of wood will frequently receive scratches over a period of time. These can usually be touched up with a special finishing stick. Resembling a crayon with a flat point, the finishing stick is rubbed along and across the scratch. When the mark is filled, wipe off any excess material with a clean cloth and buff the area slightly with the cloth. Experienced handymen will sometimes melt some of the stick and let the soft "wax" fill the damage, rather than apply it directly from the stick. This fills the mark better than just rubbing. These sticks are available in a wide assortment of colors to match most finishes and they can be used in combination to match unusual shades. If the color of the finish can't be matched exactly, try filling with a slightly darker shade than that of the wood. The result will be less noticeable than if a lighter shade is applied.

SYNTHETIC WOOD

Along with the increased cost of wood over the past several years, there have been several simulated and synthetic wood products introduced at the retail level. The most common items are made from plastic, vinyl or foam.

Vinyl comes in many forms. Perhaps the most convenient of these is rolls. If an item made of wood or metal is badly worn and not worth refinishing, it can be restored to use by covering it with a simulated wood wrapping of vinyl. The covering can be glued to the original surface or tacked underneath. For convenience, many rolls already come with an adhesive back. Vinyl has the additional advantage of being washable and moisture resistant. However, it doesn't provide much support strength so it can only be used to cover an item that is physically sound. Of course the simulated wood finish lacks the warmth and texture of real wood.

Simulated wood vinyl also comes in sheets or panels. These can be used over wood to form new surfaces for counters, furniture or cabinets and they can even be used as paneling. They are also washable and moisture resistant, but they aren't as strong as fine wood paneling. Care must be taken with the use of any simulated wood vinyl product. Unlike real wood, the pattern will come off when exposed to harsh detergents or cleaning agents.

One synthetic product that is visually difficult to distinguish from the real wood is foam. Items made of foam are lightweight and extremely easy to handle but offer little support strength. Although the texture of foam products is unlike that of real wood, they are generally used as trim in locations where they can't be too closely inspected. A typical use is providing a wood beamed effect for ceilings.

The use of imitation wood made from plastic has become increased popularity. It's frequently used as trim on furniture, in cars and in the home as door and window trim and ornaments. Unlike vinyl rolls and foam, plastic can be made with enough strength to support weight. When produced carefully, plastic can closely approximate the patterns found in real wood. However, the irregularities, texture and warmth that real wood possesses are lacking in plastic products.

REFINISHING TECHNIQUES

That old chair or dresser that you were ready to throw out can be made useful and attractive again through the proper use of refinishing procedures. Almost any wood item can be refinished as long as it is sound. Because old wood has lost some of its original moisture and isn't as pliable as it once was, it's important to prepare the wood before you attempt to refinish. Minor repairs such as the filling of holes or cracks should be made since applying new finish over damaged wood won't give satisfactory results. While some finishes can be applied directly over old ones, best results for restoring wood pieces are obtained by stripping the old finish and removing any traces down to the bare wood.

HOW TO STRIP OLD WOOD FINISHES

One technique for removing any finish is to sand, sand and sand again. Each sanding removes a little more of the old finish, but the process is extremely tedious. Dust will accumulate and a mask should be worn to give you particles. To simplify the job, there are several quality chemical strippers that you can use. They don't eliminate all the work but they do speed the stripping process. The technique for using them is as follows:

—Make sure the wood is as free of dirt and grease as possible. And that the room is well ventilated.
—Brush or pour on the stripping chemicals according to the directions on the can.
—Most stripping products must stay on the old finish for a time in order to loosen the finish. After the recommended time has elapsed, scrape off the old finish. Because the removing agents also temporarily soften the wood, it's important to scrape gently to avoid damaging the item.
—If most of the finish hasn't been removed repeat the stripping process. After coating the wood with the removing agent cover the item with waxpaper. This concentrates the strength of the solvent and slows its evaporation. To assist in removing old varnish, apply some woodshavings or rough sawdust over the surface after the solvent has been applied. Then wipe with a dry cloth.
—To remove old finish from carved areas or corners and other difficult to reach areas, wrap a wooden match stick—minus the head of course—with fine steel wool and use this self-made tool to scrape the finish away. Never use a match that could ignite because stripping agents are usually flammable.
—Allow the surface to dry and then sand gently.

Once all traces of old finish have been removed, sand the wood again to remove any small scratches that might have been made. Be sure to remove any dust from the wood and work area before applying a new finish.

Visible hardware should be removed from furniture items before the application of a new finish. Remove drawers, handles, hinges and so forth before applying any finish to the wood. Finish these items, such as draws, at the same time and in the same way that the major piece is completed. This will ensure correct color matching. When each piece has been completed re-install all the hardware and re-insert drawers or other assemblies.

PROFESSIONAL SECRETS FOR THE BEGINNER

Professional carpenters are skilled craftsmen. Many years are required for an individual to fully develop professional talents. During this time, many techniques are learned which aren't generally known to the average person. These tricks of the trade frequently simplify the job or enhance the beauty of the final product. Not all techniques work equally well for everyone. You can try different ideas and select those that work the best for you. Part of the fun and challenge of carpentry is to modify these pr ctices to meet individual needs. Don't be afraid to experiment.

Tip One. The importance of sanding before you do the final wood finishing has been stressed. However, there are times when this isn't as easy as it looks. It's often difficult to sand curved areas such as table legs or railings or to get into small crevices such as carved areas. Sanding becomes considerably easier if the sandpaper is cut into thin strips. Hold one end of a strip in each hand and pass it back and forth over curved surfaces in much the same manner that would be used to polish a pair of shoes with a shine cloth. For carved areas, work the edges into grooves and gently sand. Very ornate work requires a light touch and patience.

Tip Two. Windows will frequently stick after they have been painted. Usually this is because too much paint was used on the sides of the sash and inside the frame. Make sure that the brush you use doesn't have excessive paint on it when these areas are painted and you can avoid the problem. It will save you time if you take care when painting these areas than if you had to free sticky windows later. If you have casement windows paint the top and bottom edges first. Avoid getting paint into window hardware.

Tip Three. Installing a wall takes effort. Even an excellent installation job can appear unsatisfactory if it isn't finished with a good painting job. One mistake that's frequently made is to use a roller to get into corners and joints. This can cause roller marks or a thick application of paint in one area. Another error is the use of a roller along the large area of a wall and then using a brush to paint the corners. Professional painters know that the best results are obtained by first applying paint to corners with a brush or special corner roller and then painting the rest of the surface.

Tip Four. Dust and dirt can ruin a paint job even if you use the finest quality paint. One area where dirt frequently accumulates is along baseboards. Make sure that all traces of wax are removed before painting these areas and sand if necessary. Then use an old or cheap paintbrush to remove dust and dirt.

Tip Five. Redwood is one of the most beautiful woods and is frequently used for outdoor work. Because of its beauty and durability it is often left untreated. As it is exposed to weather, redwood assumes a silvery color. Unfortunately, the wood can also develop mildew and become darkened. This can be prevented by applying a clear, water repellent. Pigmented oil stains can also be applied to produce a special color.

Tip Six. One of the problems with restoring old furniture is that some parts might be missing or broken beyond repair. One of the difficulties is to identify the right wood to use to recreate the missing or damaged item. Identifying new wood requires some practice. Classifying wood that has been stained, painted or aged can be very tricky. However, with the wide assortment of finishing stains available even a non-matching component can be made to resemble the real thing. Best results are obtained by selecting a wood with a grain that matches the original as closely as possible. This is especially true when adding wood to a damaged part as might be the case in repairing wood with broken corners. Don't follow the edges of the wood to make the repair. Follow the grain instead. Remember, joints and repair lines that follow the direction of the grain will blend into it and be less noticeable. Lines crossing the grain direction will be emphasized.

Tip Seven. Sometimes a piece of wood furniture will become wobbly when legs wear unevenly or are broken or you might have made a slight miscalculation while you were building the item. When professionals make mistakes they don't discard the item—they correct the situation. Sometimes legs can be cut or sanded to equal lengths. If this isn't practical, it is possible to make them longer. Cut squares of matching wood and drill and bore them so that they fit over the ends of the legs. Twist them on using a good glue and then plug the holes with dowels. Cut the squres to shape with a chisel or rasp and then sand smooth. If the color of the added squares doesn't closely match the rest of the piece or if the seam shows too much don't despair. Emphasize it and make it part of the design by making a groove using rasp.

Tip Eight. Sometimes damage to wood results in an item showing a dent. These marks are not repaired by filling since the dented areas will not hold the new material. A useful trick for repairing dented areas is to cover the region with pin pricks and then soaking it with hot water. This will soften the wood fibers and help bring the wood to a uniform level. If the results aren't satisfactory, cover the area with aluminum foil and lightly press a hot iron to the dampened dent.

Tip Nine. If an item is scratched, but the marring isn't too deep, it can probably be hidden. Use a sharp utility knife to smooth the edges of the damage. The idea is to create a flat dish effect. Use a rasp and sandpaper to blend the area into the surrounding region and to smooth it. If this causes the finish to become lighter or exposes bare wood, match the repaired area to the rest of the wood by carefully staining it.

Tip Ten. The smoother an item is before it is finished the better the final result. To obtain excellent hand sanding, take a piece of an old broom handle and cover it with felt. Hold a piece of sandpaper firmly against the felt and sand the wood. Difficult ends can be smoothed effectively with this technique and it is ideal for sanding curved wood.

Tip Eleven. Best sanding results are obtained by working in the direction of the grain. Depending on the item you are sanding this might not always be possible. If sanding across the grain can't be avoided, use fine steel wool instead of sandpaper. The wood will be smoothed, but the fine marks left by sandpaper will be avoided.

Tip Twelve. The secret to applying a fine wood finish is to sand several times—each time using a finer grade of sandpaper. Eventually a point is reached where sanding produces only a fine fuzz. Additional sanding will not produce increased benefits. For most finishing applications this is far enough. If additional smoothness with superior finishing results is desired, try using a sanding sealer. These harden and stiffen the tiny wood fibers and give the wood a glass-like quality. Sanding sealers are available at paint stores or can be made at home by diluting one part of shellac with three parts of alcohol. To use it, simply brush it on smoothly and evenly and let it dry.

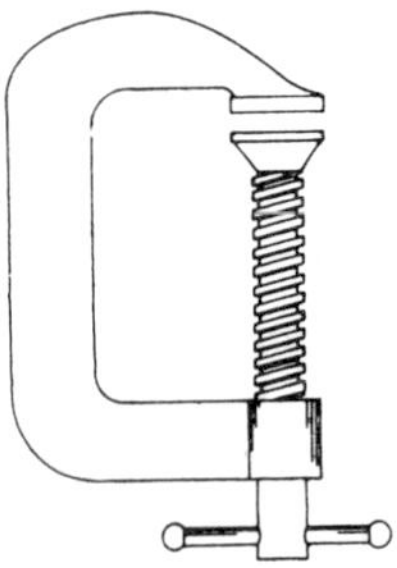

Glossary

aliphatic resin glue, Wilhold's. A strong all-purpose, all-weather glue. Ideal for porous surfaces.

alpha miracle glue, vigor aron. Sets up fast with great strength. Will not work on porous materials.

anchor. Special hardware designed to fasten together timbers or masonry.

anchor bolts. A rod-like piece of hardware used to fasten columns, girders and other construction items to masonry.

backsaw. Allows accurate cutting of straight lines. Use this tool for finishing work.

baluster. A type of small column used to support a rail when several balusters are connected by a rail they are referred to as a *balustrade*.

batten. A wood strip, usually narrow, used to fasten several pieces of wood together.

beam. A piece of heavy wood used in construction and generally thought of as an item used to support ceilings, roofs or floors. Actually it is a broad term for joists, girders and rafters.

bevel. A cut made at an angle, or a surface or edge that is prepared at an angle. Also a tool that has a rule and a moveable arm and is used to mark angles.

board. Wood that is less than 2 inches thick.

board foot. A unit of measurement for wood. It's the equivalent of a board that's 1 foot square and 1 inch thick.

bracket. A support that's used for a shelf or other item. It generally projects from a wall.

building paper. A thick paper that's used to insulate a building or addition before siding is installed. It's also occassionally placed between double floors.

build-up. A term referring to an item that has been formed from several smaller one that have been fastened together.

carpentry. The art or craft of building or repairing wooden items.

carriage bolts. This type of bolt is intended for wood-to-wood use but can also be used for fastening wood-to-metal. Use them in conjunction with a nut and a flat washer. They have roundheads and should not be driven. They are threaded only along a portion of their shafts.

carriages. An inclusive term for the supports, steps and risers for a flight of stairs.

casement. A type of window equipped with a sash that opens using hinges.

casing. The finished wood that's installed around a post or beam. It also refers to the finishing wood and trim around a door or window opening.

chalk line. Use for indicating straight lines and for temporary markings.

chamfer. A beveled surface made at the corner of a piece of wood.

checks. Cracks or splits in wood. They are generally caused by seasoninn.

chisel. Used to make small grooves, notches or cuts. Also useful for trimming and paring. Where space permits, use a plane for smoother results.

circular saw. Used for making straight cuts in unwieldy materials.

column. A support for roofs or ceilings that is comprised of a base, shaft and capital.

common wire nails. A type of nail used in many carpentry projects and for construction and house framing.

compass saw. Used for making cut-outs or carved cuts.

coping saw. Designed for making thin accurate cuts and for sawing tight curves.

crosscut saw. Used for sawing across the grain of wood.

cut nails. These are wedged shaped nails with the head on the large end. They are used to nail flooring because of their good holding power and the type of steel used to make them.

drill. Used for making small holes such as pilot holes for screws. Attachments are available for electric drills that enable you to make larger holes and to sand. However, better results are usually obtained by using a specific power tool (such as a sander) than by the use of an attachment.

drip. A projection from a window sill that protects the side of a house or building below it from water.

expansion bolts. These are used in conjunction with an expansion shield to hold materials that can not accept threaded bolts.

finishing nails. This type of nail has a small head and is made from finer wire than common nails. They are used for interior or exterior finishing work and cabinet making.

flashing. Material used to make the intersecting areas of a roof or other areas on the outside of a building that are exposed to the elements watertight.

framing. The wood which forvs the rough structure of a building.

fungi, wood. Microscopic plants that live in damp wood and cause mold, stain, and decay.

fundicide. A chemical that is poisonous to fungi.

furring. Strips of wood or metal applied to a wall or other surface to even it and normally to serve as a fastening base for finish material.

gambrel. A symmetrical roof with different slopes on each side.

gage. A tool to make a line parallel to the edge of a board.

girder. A main horizontal support beam.

hacksaw. Used to cut metal.

hammer (with curved claws). Used for driving nails and for removing nails.

hammer (with ripping claws). Used for driving nails and for separating pieces of wood tht have been fastened together.

hanger. A vertical support beam.

header. A joist that is generally short and over an opening or a joist into which ordinary joists are fitted.

headroom. Unfilled space between the floor and ceiling or between steps and ceiling.

interceptor trench. A trench filled with stone or gravel which intercepts excess water runoff before it reaches a building.

jamb. The wood found at the side of an opening—the side post.

joint. The area where different pieces of wood come together or the various methods of fitting pieces of wood together. Common types of joints: *butt.* Ends or edges that are squared and adjoining. *dovetail.* A joint formed by special pins which fit within a corresponding piece.*drawboard.* A type of mortise-and-tenon joint. It's made with holes that are cut so that the joint actually becomes tighter when pins are inserted.*fished.* A butt joint strengthened by wood nailed to the sides. *halved.* A joint that's made from two pieces of wood, each of which has had a section cut away so that they fit together flush. *lap.* A joint made when two pieces of wood overlap each other. *mortised.* A joint formed by cutting a hole (the mortise) in one piece of wood and fashioning a section (the tenon) from part of the second piece of wood so that the section fits into the hole. *miter.* A joint made by butting together two pieces of wood which have been cut at an angle.

joint cement. A powder that is usually mixed with water and used for joint treatment in gypsum-wallboard finish. Often called "spackle."

joist. The supporting members of a roof or floor which are 2 inches wide.

kerf. The term given to a cut made by any saw blade.

kiln dried lumber. Lumber that has been kiln dried often to a moisture content of 6 to 12 percent.

knot—natural finish. Softwood lumber, such as framing lumber are dried to a somewhat higher moisture content.

knot. In lumber, the portion of a branch or limb of tree that appears on the edge or face of the piece.

laths. Narrow strips of wood used to support plaster.

lacquer. A finish for wood made of resins dissolved in ethyl alcohol. When it dries it leaves a tough coating.

lattice. A framework of crossed wood or metal strips.

lag screws. Use these when standard wood screws would be too light or too short. They are longer and heavier than ordinary wood screws.

level. The position of a line or surface when parallel to the surface of still water. Generally thought of as synonymous with "truly horizontal." The term also refers to a tool for determining a horizontal positon.

machine bolts. This type of bolt is precision made for use in metal-to-metal applications where close tolerance is important. Heads are square, hexagon, double hexagon, rounded or flat. They are meant to be used in conjunction with a nut that is usually the same shape as the head.

mullion. The pieces of wood between window frame openings.

natural finish. A transparent finish which does not seriously alter the original color or grain of the natural wood. Natural finishes are usually provided by sealers, oils, varnishes, water-repellent preservatives, and other similar materials.

newel. The main post of the bottom of a staircase or winding staircase.

notch. A crosswise rabbet at the end of a board.

o.c., on center. The measurement of spacing for studs, rafters, joists, and the like from the center of one member to the center of the next.

oak, English. The oak used for medieval ships and furniture. Brown, open-grained, with prominent figuring when quarter-sawn. Very strong and durable.

oak, Japanese and Austrian. More mellow and easier to work than English oak. Good furniture woods.

pier. Supports, usually concrete, which are separate from the foundation of a building.

plane. Used for edge straightening or shaving wood. Use a long plane for shaving long surfaces. If the job involves cutting along the end of the wood, use a block plane.

pulley stile. The part of the window frame that contains the pulleys. The edges of the sash slide between them.

purlin. A secondary roof or floor member spanning between beams, running perpendicular to them.

putty. A type of cement usually made of whiting and boiled linseed oil, beaten or kneaded to the consistency of dough, and used in sealing glass in sash, filling small holes and crevices in wood, and for similar purposes.

quarter round. A small molding that has the cross section of a quarter circle.

rabbet. A rectangular longitudinal groove cut in the corner edge of a board or plank.

rafter. A beam that forms the main body of a building's roof and that slopes from the roof's ridge to the eaves.

ripsaw. Use this saw to cut wood in the direction of the grain.

riser. The vertical board positioned between two stair treads.

roofing nails. These are round shafted with large heads and diamond points. They are usually 1 inch or 1¼ inch in length but they can be longer. They are used to fasten asphalt shingles.

sabre saw. Used for cutting irregular or wavy shapes. Can also be used for cutting straight lines.

sash. The portion of a window's frame that holds the glass.

scaffold. A platform that helps workmen reach high places and enables them to work on the sides of a building.

scaffold nails. These nails have two heads to make it easier to remove them after they have been temporarily positioned.

scale. A short measure used in drawings and plans instead of actual distances. One-quarter of an inch on a drawing is usually equivalent to 1 foot of actual distance.

scantling. Lumber that has a cross-section which ranges from 2 by 4 inches to 4 by 4 inches.

screwdrivers (regular slot). Used for positioning screws with a single slot in the head or for removing such screws.

screwdrivers (Phillips). Used to position screws that have a + slot in the head or for removing such screws.

shakes. Imperfections in wood caused by high winds or adverse conditions while the tree it came from was still growing. Also a type of siding for a house.

sheathing. Solid wall covering which is covered with siding.

sheet metal screws. They have round, oval, flat or fillister heads and are used to fasten metal parts.

shellac. A transparent coating made by dissolving lac, a resinous secretion of the lac bug (a scale insect that thrives in tropical countries, especially India), in alcohol.

spray adhesive 54, Wilhold's. For spraying large areas. A contact-type adhesive; however, pieces may be moved if necessary. Wallpaper and veneer sheets can be mounted with this adhesive.

spredhesive. A pressure-sensitive, acrylic base adhesive. Takes the place of double face adhesive tape.

stainless glue. Excellent for gluing veneers and inlays, as well as other woodworking assemblies.

siding. The finish wall covering.

silicone glue. Used for insulating exposed electrical wires and adhering lamp parts. Remains flexible. Good for filling cavities.

sobo white glue. A good all-purpose glue. Similar to other white glues. Dries crystal clear.

square. A tool used for accurate measurements and for determining right angles.

stapler. Used for speed fastening instead of nailing or tacking.

stove bolts. These have either round or flat slotted heads with threads along the entire shaft. They are used in conjunction with square nuts in metal-to-metal, wood-to-metal or wood-to-wood applications where precision is not a critical factor. Flathead stove bolts should be countersunk. Roundhead types should be tightened until they are flush with the surface.

stud. One of a series of slender wood or metal vertical structural members placed as supporting elements in walls and partitions (Plural: studs or studding.)

subfloor. A wooden floor placed over the floor joists. The finished floor is laid over it.

threshold. A curved or beveled item, generally wood, over which a door swings.

timber. Wood with a cross-section greater than 4 by 6 inches. Examples are posts, sills and girders.

top plate. The wood that supports the ends of rafters.

tread. That part of a step which is horizontal.

undercoat. A coating applied prior to the finishing or top coats of a paint job. It may be the first of two or the second of three coats. In some usage of the word it may become synonymous with priming coat.

under layment. A material placed under finish coverings to provide a smooth, even surface for applying the finish.

valleys. The interior angle formed by the slopes of two intersecting roofing surfaces.

wainscoting. Panel work which covers the lower portion of a wall.

wallpaper adhesive. One of the best is vinyl wall covering paste, which is a mixture of wheat flour and a special adhesive (Staz-Tite Products).

wall sizing. Porous wall surfaces must be sized before wallpaper is applied. One very good size is Wal-Treet, an acrylic base coat (Paint Manufacturers, Inc.). Shellac, flat paint, undercoater, and other materials may also be used.

wood screws. Use these instead of nails for extra holding power. They are designed according to the type of head—flat, oval or round. Use the flathead or ovalhead screws with countersinking to hide the screw beneath filler or a covering material. Roundhead screws should not be countersunk. Tighten them until the heads are flush with the surface.

wrench. Use to assist in turning hardware. Especially useful for tightening or removing nuts and large screws.

yellow birch. Yellow birch is heavy and hard and is therefore difficult to work with hand tools. It is used principally for lumber, veneer and crossties. Because of its pleasing grain pattern and ability to take a high polish, it is widely used for cabinets and furniture. The heartwood is light reddish-brown.

yellow-poplar. Yellow-poplar grows in many eastern and southern states. It is moderately light in weight and has an excellent reputation for taking and holding paint. It stains and glues quite satisfactorily. It does not impart taste or odor to foods and therefore is used for food containers. Other uses are lumber, veneer and pulpwood. It is brownish yellow with a greenish tinge.

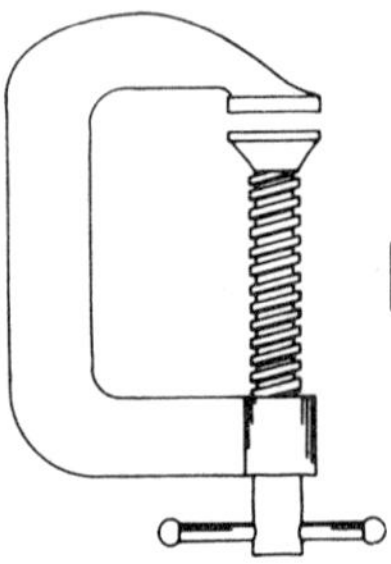

Index